Robin C. Kirby

The Topology of 4-Manifolds

Springer-Verlag
Berlin Heidelberg New York
London Paris Tokyo
Hong Kong Barcelona
Budapest

Author

Robin C. Kirby
Department of Mathematics
University of California
Berkeley, CA 94720, USA

This volume belongs to the Subseries:
Nankai Institute of Mathematics
Tianjin, P. R. China (vol. 6) of the LNM

Advisers: S. S. Chern, B.-j. Jiang

1st Edition 1989
2nd Printing 1991

Mathematics Subject Classification (1980): 57M99, 57N13

ISBN 3-540-51148-2 Springer-Verlag Berlin Heidelberg New York
ISBN 0-387-51148-2 Springer-Verlag New York Berlin Heidelberg

Printed in Germany

Printing and binding: Druckhaus Beltz, Hemsbach/Bergstr.
46/3140-543210 - Printed on acid-free paper

PREFACE

In the late 1970's, Mike Freedman and I sketched an argument using immersion theory for showing that $\Omega_4^{SO} = Z$. In 1982–83, Iain Aitchison and I worked out new proofs and a reorganization of $\Omega_4^{\text{spin}} = Z$, $p_1 = 3\sigma$, and Rohlin's theorem. In the last 5 years, further simplifications including a yet easier proof of $\Omega_4^{\text{spin}} = Z$ have been found.

A first draft of Chapters XII and XIII was written at IMPA in Rio de Janeiro in fall 1982 and other bits at the University of Maryland in spring 1983, but the bulk of the writing was done at S.-S. Chern's suggestion at the Nankai Institute of Mathematics in May 1987. I was very ably assisted by Bao–zhen Yu, who found some gaps and corrected many errors, not all minor. I am indebted to Charles Livingston and the topology seminar at Indiana who found further gaffes in Fall 1987, and to Berkeley students, particularly Chris Herald, for checking the final version.

Recent work with Larry Taylor on Pin structures and non-orientable generalizations of Rohlin's Theorem has fed back into some further sharpenings of Chapter IV and the proof of Rohlin's Theorem.

Thanks to my collaborators, to IMPA, Maryland, and especially Nankai for their warm hospitality, to Faye Yeager for an excellent TeX manuscript, and to Deb Craig for help with the many figures.

TABLE OF CONTENTS

INTRODUCTION

When I began to think about 4-manifolds in 1973, the basic theorems included the Whitehead–Milnor theorem on homotopy type [**Wh**], [**Milnor1**], Rohlin's Theorem [**Rohlin**], $\Omega_4^{SO} = Z$, $\Omega_4^{\text{spin}} = Z$, the Hirzebruch index theorem $p_1 = 3\sigma$, and Wall's theorems on diffeomorphisms and h-cobordism [**Wall1**] and [**Wall2**]. These theorems were untranslated ([**Rohlin**]) or unreadable ([**Wh**]), or were special cases of big machines in algebraic topology ($\Omega_4^{SO} = Z$, $\Omega_4^{\text{spin}} = Z$, $p_1 = 3\sigma$), or, even though accessible, could, with hindsight, use streamlining (Wall's theorems).

In the early 1970's, Casson and Rohlin independently gave geometric proofs of Rohlin's Theorem and improvements followed ([**F-K**] and Y. Matsumoto and Guillou and Marin in [**G-M**]). Rohlin's proof of $\Omega_4^{SO} = Z$ was translated [**G-M**] and lectured on by Morgan and others, with the latest version in [**Melvin**]. But a geometric, low dimensional proof of Ω_4^{spin} was missing. The algebraic topological proofs are powerful, and beautiful mathematics in their own right, but there ought to be proofs of the fundamental 4-manifold theorems which belong to the field of 4-dimensions (or less), and prepare the student in the geometric side of the theory.

We give a geometric proof of $\Omega_4^{SO} = Z$ starting with an immersion of M^4 into R^6; it is different but not necessarily better than the proofs mentioned in the previous paragraph. It's unique virtue was that Iain Aitchison and I were able to make it work for Ω_4^{spin}, but not without some difficulties. Recently, a simple proof of $\Omega_4^{\text{spin}} = Z$ turned up, which only uses the fact (not the method of proof) that M^4 bounds if $p_1(M) = 0$. This work led to an improved proof of Rohlin's theorem using spin structures. These proofs are first presented here. Handlebody theory is also exploited to streamline some proofs, e.g., Wall's theorems, and a few new wrinkles are included here and there.

Chapters XII–XIII give a sketch of Casson's and Freedman's work on topologic handles and 4-manifolds. These chapters might profitably be read as an introduction to Freedman's fundamental paper [**Freedman1**] or concurrently with Casson's 1974 notes in [**G-M**]. Chapter XIV contains constructions of exotic smooth structures on R^4, a countable number which do not imbed in S^4 and one that does imbed in S^4.

A reader needs a good, intuitive understanding of smooth manifolds and bundles, knowledge of the simplest form of the immersion theorem (perhaps best read in [**H-P**]), and a decent understanding of characteristic classes as applied to low dimensions using the obstruction theory definition [**M-S**, chapter 12].

Framed links are used as the basic way of describing 4-manifolds; Chapter I covers this material. Homotopy type, intersection forms, characteristic classes and the index fall in Chapter II. Chapter III states classification theorems as of July 1987.

Spin structures are tricky fellows, especially over S^1 and surfaces, and they are presented carefully, I hope, in Chapter IV, with a fundamental example in V. Chapters VI–IX focus on the proofs that $\Omega_4^{SO} = Z$, $\Omega_4^{\text{spin}} = Z$, and $p_1 = 3\sigma$, beginning with the study of immersions and singular sets in VI. The remaining chapter titles are self explanatory.

II.3.1 refers to Theorem or Lemma 1 in §3 of Chapter II; 3.1 refers to Theorem or Lemma 1 in §3 of the same chapter. Similarly with figures. □ marks the end of a proof.

I. HANDLEBODIES AND FRAMED LINKS

§1. Handlebodies.

A handlebody decomposition of a compact manifold M^m is a sequence $B^m = M_0 \subset M_1 \subset M_2 \subset \cdots \subset M_k = M$ where M_i is obtained from M_{i-1} by adding a k_i-handle, that is, $M_i = M_{i-1} \underset{f_i}{\cup} B^{k_i} \times B^{m-k_i}$ where $f_i : \partial B^{k_i} \times B^{m-k_i} \to \partial M_{i-1}$ is an imbedding which is called the attaching map (Figure 1.1). $M_0 = B^m = B^0 \times B^m$ is a zero-handle and there may be others. Handlebody decompositions exist for the categories TOP, PL and DIFF except for the case of 4-dimensional topological manifolds which are handlebodies iff they are smoothable (see [**K-S**] and [**Quinn**]). We are only interested in the smooth case where f_i has to be a smooth imbedding. Then M_i has "corners" where the k_i-handle was attached (Figure 1.1), but the phrase "corners can be smoothed" has been a phrase that I have heard for 30 years, and this is not the place to explain it.

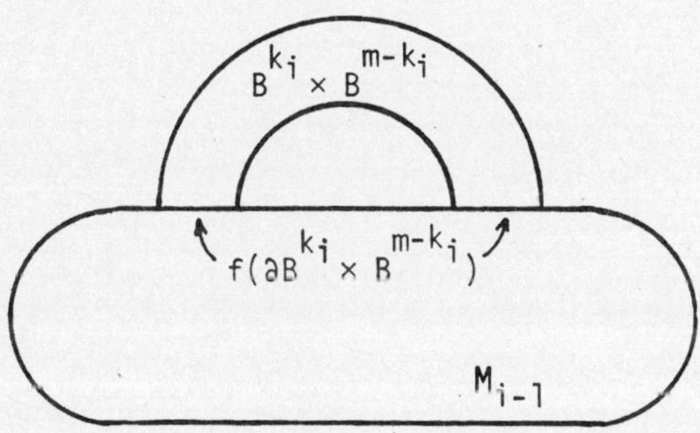

Figure 1.1

Smooth handlebody decompositions (handlebodies for short) correspond to Morse functions $h : M \to R$ (which have non-degenerate critical points at different levels). A critical point of h corresponds to $0 \times 0 \in B^{k_i} \times B^{m-k_i}$ and $B^{k_i} \times 0$ is the descending manifold and $0 \times B^{m-k_i}$ is the ascending manifold.

According to [**Cerf1**], any two Morse functions h_0, h_1 are homotopic by an arc h_t of functions, $t \in [0,1]$, which are Morse functions for all but finitely many t, at which h_t either has two critical points at the same level or a birth or a death occurs.

A death corresponds to a pair of handles cancelling and a birth to the creation of a pair, as is shown in Figure 1.2.

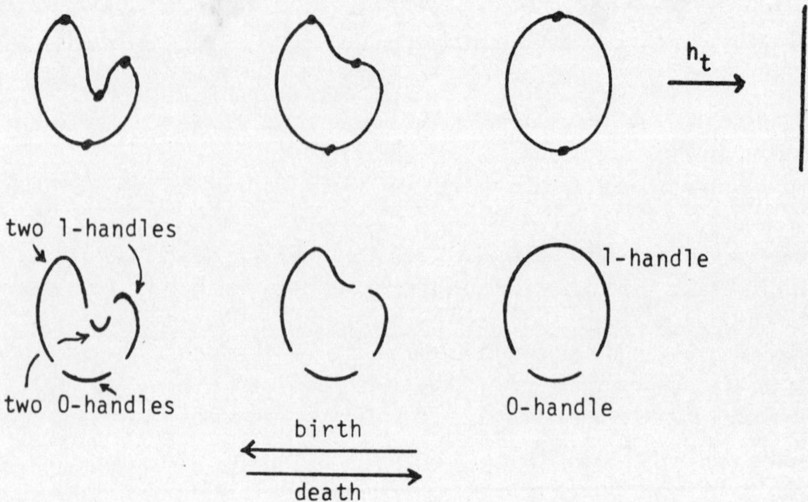

Figure 1.2

Thus, as we homotop h_0 to h_1, the h_t move through Morse functions, which correspond to isotopy of the attaching maps f_i, and whenever a birth or death is passed, a pair of handles are either created or cancelled.

We can summarize this by

THEOREM 1.1. *Any two smooth handlebody decompositions of M^m are related by isotopy of the attaching maps and creation or cancellation of handle pairs.*

It should be noted that handles can always be attached in the order of their indices. For if a $(k+1)$-handle $B^{k+1} \times B^{m-k-1}$ is attached first and then a k-handle $B^k \times B^{m-k}$, then by transversality the attaching sphere of the k-handle, $S^{k-1} \times 0$ misses the cosphere of the $(k+1)$-handle, $0 \times S^{m-k-2}$, (since $k-1+m-k-2 < m-1$) and hence can be isotoped off of the $(k+1)$-handle and added first. Moreover, the same argument shows that two k-handles can be attached in either order.

§2. Framed Links.

In dimension 4 we will visualize handlebodies by drawing their attaching maps, when possible, in $\partial M_0 = \partial B^4 = S^3$.

A 1-handle is attached by $S^0 \times B^3$, so we draw a pair of 3-balls in S^3 as in Figure 2.1. Often it will be convenient to denote a 1-handle by an unknotted circle with a "dot" on it. The circle bounds an obvious disk, and if we push that disk into B^4 (so that $(B^2, S^1) \to (B^4, S^3)$ is a proper imbedding) and remove a neighborhood of it, then the remainder is $S^1 \times B^3$, the result of adding a 1-handle to B^4. Thus arcs that go over the 1-handle should be drawn so as to go through the dotted circle.

Figure 2.1

We draw the attaching map of a 2-handle, $f(S^1 \times B^2)$, by drawing $f(S^1 \times 0)$, a knot in S^3, and labeling the knot with an integer, its framing. Let F^2 be a surface in B^4 which $f(S^1 \times 0)$ bounds. Then $f(S^1 \times B^2)$ corresponds to the zero framing of $f(S^1 \times 0)$ if it is the trivialization of the normal bundle of $f(S^1 \times 0)$ which extends to the normal bundle of F^2 in B^4. Equivalently, let F^2 be a Seifert surface for $f(S^1 \times 0)$ in S^3; then the zero-framing is the one for which $f(S^1 \times (1,0))$ is tangent to F^2. Framing k means that $f(S^1 \times B^2)$ differs from the zero framing by k full twists around $f(S^1 \times 0)$ (right-handed for $k > 0$, left for $k < 0$), that is, by $k \in Z = \pi_1(SO(2))$.

Figure 2.2 gives some examples where we have drawn $f(S^1 \times 0)$ and $f(S^1 \times e_1)$ for $e_1 = (1,0) \in B^2$.

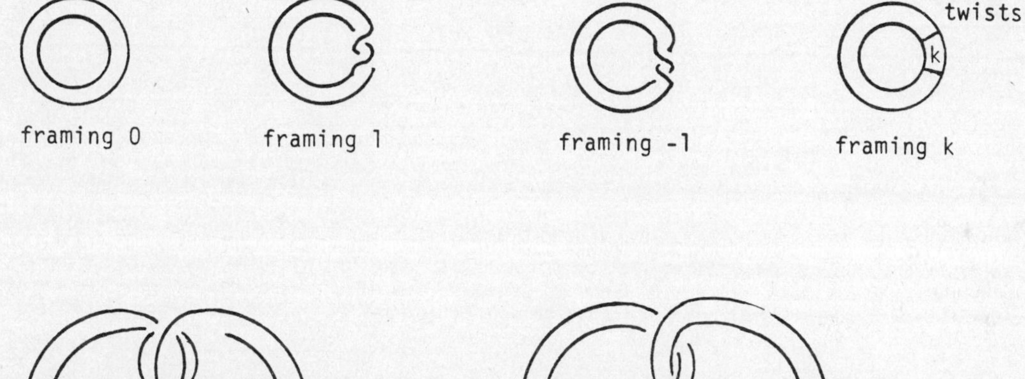

Figure 2.2

Sometimes the attaching circle of a 2-handle goes over a 1-handle; it is drawn as in Figure 2.3.

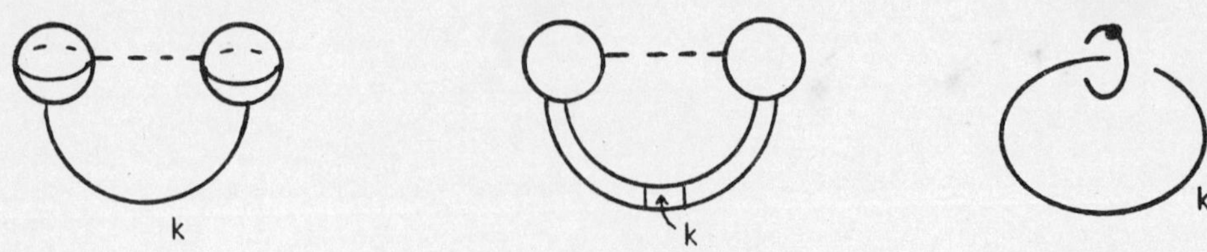

Figure 2.3

Then the attaching circle does not bound a Seifert surface in B^4, so to describe the framing we could draw $f(S^1 \times e_1)$. However, it is more convenient to fix a dotted line joining the two feet of the 1-handle and then to assume that $f(S^1 \times 0)$ goes parallel to the dotted line rather than over the 1-handle; now $f(S^1 \times 0)$ has a Seifert surface and a well defined zero framing. One has to be careful, when isotoping attaching maps, not to cross the dotted line, for that changes the zero-framing just as it would if we changed a crossing in $f(S^1 \times 0)$ (see Figure 2.4).

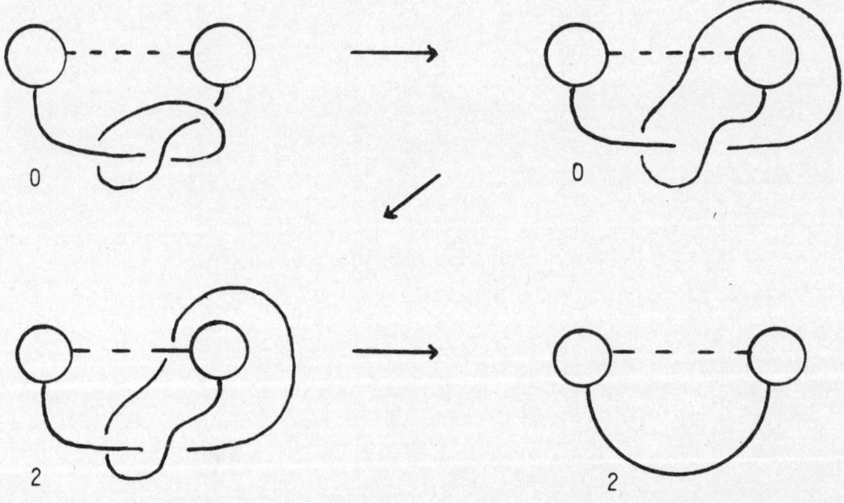

Figure 2.4

If $f(S^1 \times 0)$ goes algebraically zero times over the 1-handle, then it has a Seifert surface and the framing is defined without the use of a dotted line.

When we switch notation for the 1-handle to the circle with a dot, then we place the dotted circle so as to link the dotted line, and draw all the 2-handle attaching circles parallel to the dotted line through the dotted circle (Figure 2.3).

Adding a 1-handle to B^4 results in $S^1 \times B^3$ with boundary $S^1 \times S^2$. Adding a 2-handle to an unknot with zero framing gives $S^2 \times B^2$, also with boundary $S^1 \times S^2$. Handles which are attached later cannot tell what the $S^1 \times S^2$ is the boundary of. Switching the 1-handle to the 2-handle is the same as doing surgery on the obvious S^1 defined by the 1-handle with the trivial framing. The next lemma follows as an exercise from this discussion.

LEMMA 2.1. *Surgery on the S^1 defined by a 1-handle corresponds to removing the dot from the dotted circle and replacing it with a zero (if the trivial framing of the normal bundle of S^1 was used for surgery).*

If there are no 1-handles, then there is an obvious linking matrix A associated with the 2-handles: a_{ij} is the linking number of the ith and j^{th} attaching circles which are oriented by the standard counterclockwise orientation of ∂B^2. a_{ii} is just the framing of the ith handle. A is symmetric, and later will be seen to be the intersection matrix on the second homology of the 4-manifold (II, §1).

If there are 1-handles, we can draw them as dotted circles (oriented arbitrarily) and form an extended linking matrix A' where a_{ii} for a dotted circle is defined to be zero (as if surgery on the 1-handle was performed) and a_{ij} for a 1- and 2-handle is just the algebraic number of times the 2-handle goes over the 1-handle (or the linking between the dotted circle and the attaching circle). Two 1-handles must always be geometrically unlinked. So the extended linking matrix A' has the form

$$\begin{matrix} \text{1-handles} \{ \\ \text{2-handles} \{ \end{matrix} \left(\begin{array}{c|c} 0 & * \\ \hline * & * \end{array} \right) = A'.$$

3-handles are attached by an imbedding $f : S^2 \times B^1 \to \partial M_i$. The framing is uninteresting, but 2-spheres are hard to draw, especially non-trivial ones. (A complicated one is drawn in [**H-K-K**], §4.)

However, the 3-handles and 4-handle of a closed M^4 together are diffeomorphic to $\overset{k}{\natural} S^1 \times B^3$ (a 0-handle and k 1-handles), with boundary $\overset{k}{\sharp} S^1 \times S^2$. So the 3- and 4-handles are attached by a diffeomorphism of $\overset{k}{\sharp}(S^1 \times S^2)$. But any such diffeomorphism extends over $\overset{k}{\natural}(S^1 \times B^3)$ [**L-P**], so it makes no difference how the 3- and 4-handles are attached. For the case $\partial M \neq \emptyset$, [**Trace2**] gives useful information on attaching 2-handles.

Given a framed link L, perhaps containing dotted circles, let M_L^4 denote the 4-manifold obtained by adding handles to the link L. This is a smooth 4-manifold with boundary ∂M_L. However, if ∂M_L is S^3 or $\overset{k}{\sharp} S^1 \times S^2$, then we can close up M_L by adding a 4-handle and perhaps 3-handles. In this case M_L may refer to either the manifold with boundary or the closed 4-manifold, according to context.

Given L, it is useful to know how to describe the double of M_L along ∂M_L. In this case, the 0-handle generates a 4-handle, and 1-handles generate 3-handles, and each 2-handle generates another 2-handle which is added to the co-circle, $0 \times \partial B^2$, of the generating 2-handle. This co-circle gets a framing from its neighboring co-circles $* \times \partial B^2$ which do not link ∂B^2, so the framing is zero. Thus we have shown

LEMMA 2.2. *Given M_L, the framed link for the double DM_L is obtained by adding unknotted circles, linking each 2-handle geometrically once, with framing zero as in Figure 2.5.*

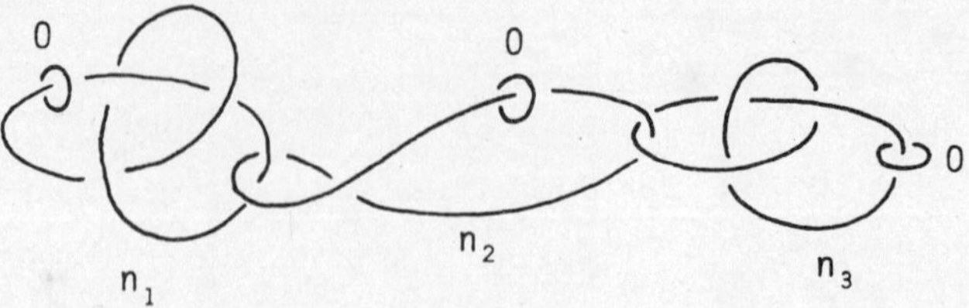

Figure 2.5

§3. Examples.

Figure 3.1

The 4-sphere is the empty link and $S^2 \times S^2$, $S^2 \tilde{\times} S^2$ and $\pm CP^2$ are drawn in Figure 3.1. To see these more clearly, note that the B^2-bundle ξ_k over S^2 with Euler class k (actually k times the generator of $H^2(S^2; Z)$) can be described by adding a 2-handle $B^2 \times B^2$ to an unknot in B^4 (thought of as $\partial B^2 \times 0$ in $B^2 \times B^2 = B^4$) with framing k; the S^2 is $B^2 \times 0 \cup B^2 \times 0$ and the framing gives the twist in ξ_k. Then $\pm CP^2$ is $\pm\xi_1 = \xi_{\pm 1}$ with a 4-handle attached to $\partial\xi_{\pm 1} = S^3$. The non-trivial S^2-bundle over S^2, $S^2 \tilde{\times} S^2$, has a fiber with trivial normal bundle and a section with non-trivial normal bundle (the left and right components of the link).

For a really non-trivial example of a 1-connected, closed M_L (it is trivial to draw non-closed examples—any link will do—but rarely is the boundary equal to $\overset{k}{\sharp} S^1 \times S^2$), we must turn to the Kummer surface. It is a complex surface with many definitions of the underlying 4-manifold, e.g., any nonsingular quartic in CP^3, say $x^4 + y^4 + z^4 + w^4 = 0$, (see [**H-K-K**]). Figure 3.2 shows a framed link for it with no 1- or 3-handles. It consists of a trefoil knot with framing zero and a small linking circle with framing -2. "On" a Seifert surface for the trefoil knot, draw twenty circles, weaving as drawn, all with framing -2. These twenty-two 2-handles, with a 0- and a 4-handle, describe the Kummer surface.

Our examples do not require 1- and 3-handles. It is not known whether a simply connected, closed 4-manifold needs 1 and/or 3-handles, but the Dolgachev surface ([**H-K-K**] §§3,4 and [**Don3**]) is a good candidate for needing them.

If $\partial M \neq \emptyset$, then simply connected 4-manifolds may require 1 or 3-handles; for example any contractible 4-manifold other than B^4 must have 1 or 3-handles. Casson gave a construction that produces contractible 4-manifolds that need 1-handles specifically ([**Kirby2**] Problem 4.18). Suppose that a contractible M^4 can be made without 1-handles. Then, inverting the Morse function, M^4 can be constructed from ∂M by adding the same number of 1-handles and 2-handles, and one 4-handle. It follows that $\pi_1(\partial M)$ can be killed by adding the same number of generators and relations. But a theorem of Gerstenhaber and Rothaus [**G-R**] states that a finitely presented group with

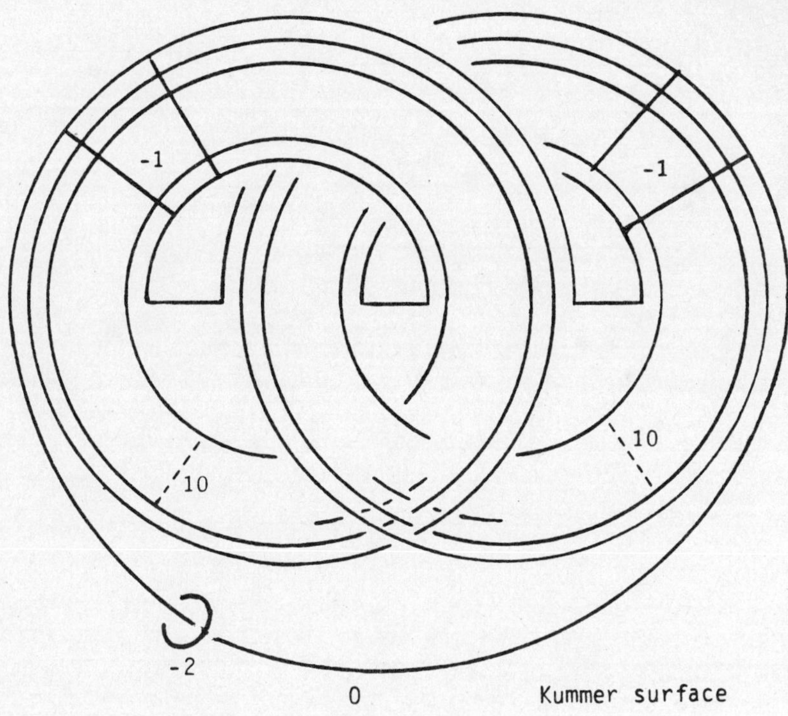

Figure 3.2

a representation into a linear group cannot be killed with an equal number of generators and relations. So any contractible M^4 whose $\pi_1(\partial M)$ has such a representation requires 1-handles. For a specific example, choose a Brieskorn homology 3-sphere which bounds a contractible 4-manifold, e.g. $\sum(2,3,13)$ (see [**A-K4**]) or $\sum(p, ps-1, ps+1)$ for p even, s odd (see [**C-H**] for other collections); note that $\pi_1(\sum(p,q,r))$ is a discrete subgroup of a compact, connected Lie group [**Milnor4**].

§4. Handle Slides.

According to Theorem 1.1 any two handle decompositions for M^4 are related by isotopy of attaching maps and births and deaths. In the language of framed links, a birth of a 1-2 handle pair or 2-3 pair is shown in Figure 4.1 by the sudden appearance, **away** from the rest of the link, of the indicated links. A death (or cancellation) is their **disappearance**.

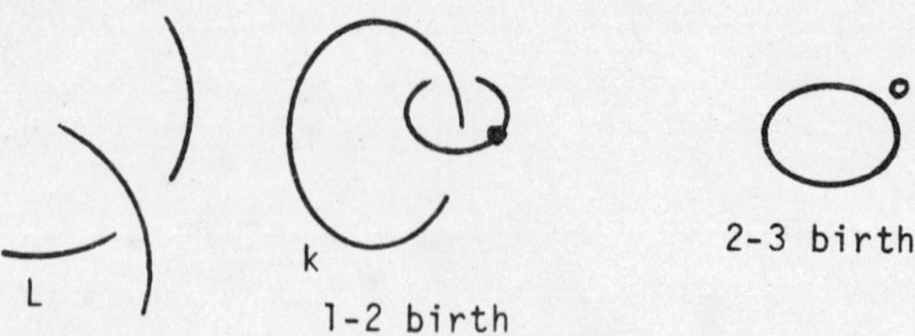

Figure 4.1

An isotopy of an attaching map becomes interesting when it goes "over" another handle rather than just moving about in $S^3 = \partial(\text{0-handle})$. The reader should picture an attaching circle which goes over the top of another 2-handle intersecting the critical point (the north pole) of the second 2-handle. If the attaching map is perturbed "left", it falls down to one side of the second attaching circle, if "right", then to the other side, Figure 4.2.

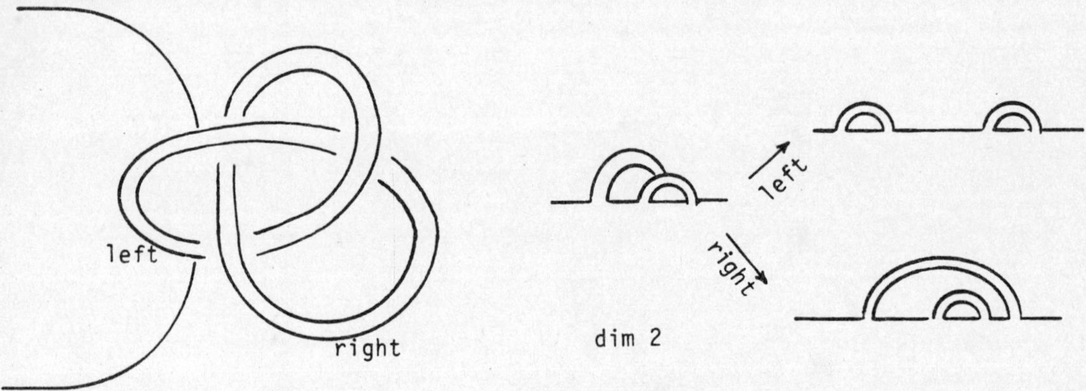

Figure 4.2

Thus the process of sliding one 2-handle over another (of going from "left to right"), is to take the band-connected sum of the first attaching map with a push-off of the second attaching map, using the framing to determine the push-off (Figure 4.3).

The band-connected sum can be done along any band, which is allowed to have any number of right or left half twists in it. The attaching circles should be oriented and then the band-connected sum will either "add" or "subtract" the push-off from the first circle.

The new framing can be computed from the linking matrix by the same process as a change of basis; if α slides over β, then the new basis should be $\alpha \pm \beta$ and β with framing and linking as in Figure 4.4. In Figure 4.3, $m = 0$. The reader can verify this by drawing $f(S^1 \times 0)$ and $f(S^1 \times e_1)$ for each handle, doing the band-connected sum, and computing the new linkings.

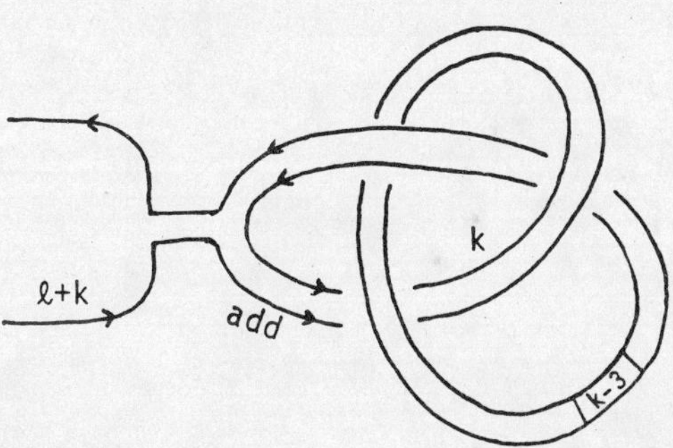

Figure 4.3

$$\begin{matrix}\alpha \\ \beta\end{matrix}\begin{pmatrix} \ell & m \\ m & k \end{pmatrix} \longrightarrow \begin{matrix}\alpha\pm\beta \\ \beta\end{matrix}\begin{pmatrix} \ell \pm 2m + k & m+k \\ m+k & k \end{pmatrix}$$

Figure 4.4

The same thing works if we "slide a 2-handle over a 1-handle"; we are thinking of the 1-handle as a dotted circle with a "framing" zero, and sliding a 2-handle over it (Figure 4.5) is the same as isotoping the attaching map of the 2-handle between the feet of the 1-handle (Figure 2.4). Note that the framing changes according to the change in the linking matrix when a 1-handle is added to a 2-handle, which corresponds to crossing the dotted line in Figure 2.4.

It is possible to make sense of sliding a 1-handle (dotted circle) over a 2-handle whose attaching circle is a slice knot ([**A-K3**], pg. 376); the knotted, dotted circle means remove the slice disk from B^4. But we won't pursue this notion, and from now on rule out the possibility of sliding a 1-handle over a 2-handle.

At this point there are a number of elementary examples that should be understood.

LEMMA 4.1. *An unknotted S^1 with framing ± 1 can always be moved away from the rest of the link L with the effect of giving all arcs going through S^1 a full ∓ 1 twist and changing the framings by adding ∓ 1 to each arc, assuming the arcs represent different components of L (in general they change according to change of basis in the linking matrix). See Figure 4.6.*

PROOF: First do the case for one arc, $k = 1$, by sliding the arc once over the circle; we add if the linking between the oriented arc and circle is ∓ 1 compared to ± 1, and subtract otherwise. In general slide all arcs over the circle once. □

COROLLARY 4.2. $S^2 \tilde{\times} S^2 = CP^2 \sharp - \hat{C}P^2$.

PROOF:

=

0 1 -1 1 □

COROLLARY 4.3. $(S^2 \times S^2) \sharp CP^2 = CP^2 \sharp (-CP^2) \sharp CP^2$.

PROOF:

= =

1 0 0 1 1 1 1 -1 1 □

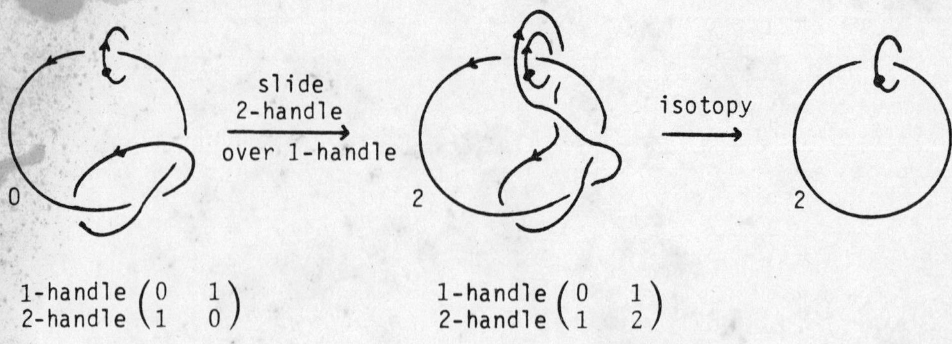

Figure 4.5

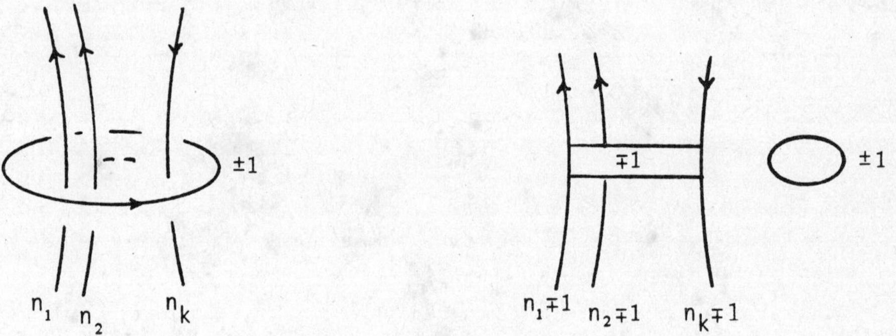

Figure 4.6

LEMMA 4.4. (2k, 0) = (0, 0) (2k+1, 0) = (1, 0)

PROOF: Each time the left circle is slid over the right (with the proper band-connected sum), the framing changes by ± 2. □

LEMMA 4.5. *If in L (with no 1-handles) a component L_0 is an unknot with framing zero which links only one other component L_1 geometrically once, then $L_0 \cup L_1$ may be moved away from the rest of L without changing framings. Then L_1 can be unknotted* *and its framing changes to* 0 *or* 1.

PROOF: If a strand of L_i crosses L_1, L_0 can be used to change the crossing without changing framings (Figure 4.7). An iteration of this move proves the first statement. The same move changes crossings of L_1 itself, thereby unknotting L_i and changing its framing by an even integer. An application of Lemma 4.4 changes the framing of L_1 to 0 or 1. □

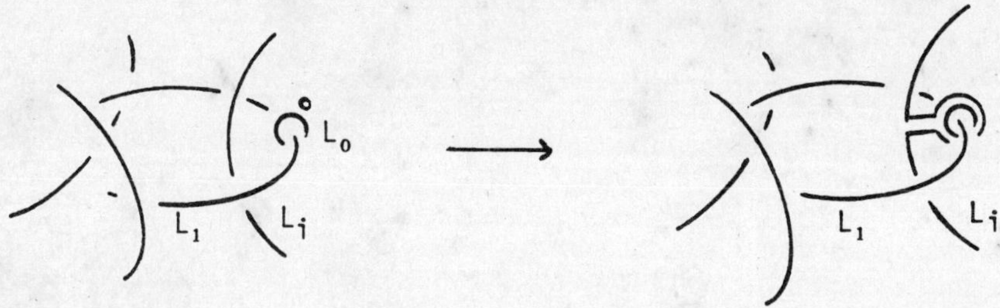

Figure 4.7

COROLLARY 4.6. *If* L *(no 1-handles) has a component* L_i *whose framing is odd, the* $M_L \sharp S^2 \times S^2 = M_L \sharp S^2 \tilde{\times} S^2$.

PROOF: As in Figure 4.8, slide L_1 over L_i, giving L_1 an odd framing. Then use L_0 as in Lemma 4.5 to free L_1 from L, unknot L_1, and change its framing to one. □

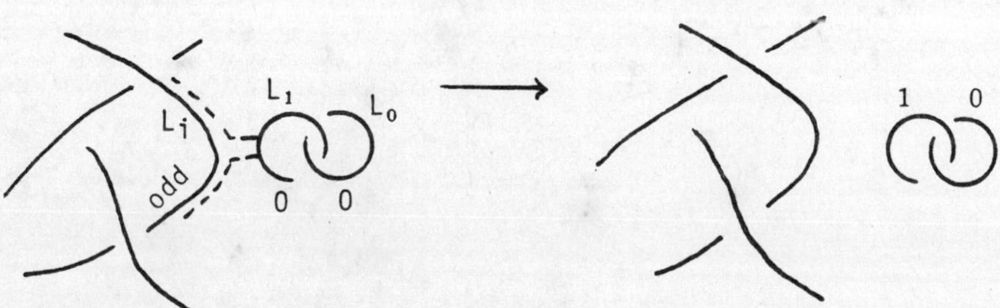

Figure 4.8

Now we re-examine the cancellation of a 1-2-handle pair. These should cancel if the 2-handle goes over the 1-handle geometrically once. If so, what happens to the other 2-handles that go over the 1-handle? Slide these 2-handles over the cancelling 2-handle so that they are free from the 1-handle (framings can change). Then isotope the cancelling pair away and erase them. The isotopy is obvious if the 1-handles are drawn with feet ($S^0 \times B^3$), for isotope one foot along the attaching map of the cancelling 2-handle; in the dotted circle notation, slide other 2-handles over the dotted circle to free the cancelling pair, as was done in Lemma 4.5.

§5. ∂M^4.

Every orientable 3-manifold N^3 can be obtained by surgery on a framed link L in S^3 [**Lickorish**] which is the same as saying that $N^3 = \partial M_L$. We show later that N bounds a spin 4-manifold M^4 (Theorem VII.3); the 1-handles of M^4 can be traded for 2-handles (Lemma 2.1) and (inverting M) the 3-handles can be traded for 2-handles. Because M is spin, the framings on the 2-handles are all even (see II §4), so we see this way that $N^3 = \partial M_L$ for L with even framings (compare [**Kaplan**]).

THEOREM 1.5.1 [**Kirby1**]. *Suppose $\partial M_{L_1} = \partial M_{L_2}$ for links L_1 and L_2. Then L_1 can be changed to L_2 by a series of the following moves:*

1) *Slide one 2-handle over another (this does not change M_L).*
2) *Add or remove an isolated copy of an unknotted circle with framing ± 1 (this changes M_L to $M_L \# \pm CP^2$ or vice versa).*

There are few applications of Theorem 5.1 other than this: if N_1^3 and N_2^3 are suspected to be diffeomorphic, then Theorem 5.1 encourages one to prove it by moves (1) and (2). A reduction to one move is given in [**F-R**].

The phrase "*blowing down*" (it comes from algebraic geometry) refers to identifying an unknotted circle in L with framing ± 1, using Lemma 4.1 to move it away from the rest of L, and then using move (2) to delete the circle. "Blowing up" is the reverse procedure.

EXAMPLE 5.2: Using Lemmas 4.1–4.5, we see that $\partial M_{L_i} = \partial M_{L_j} = S^3$ for all the links in Figure 5.1.

Figure 5.1

EXAMPLE 5.3: The 3-torus T^3 can be described as the boundary of either of the links in Figure 5.2.

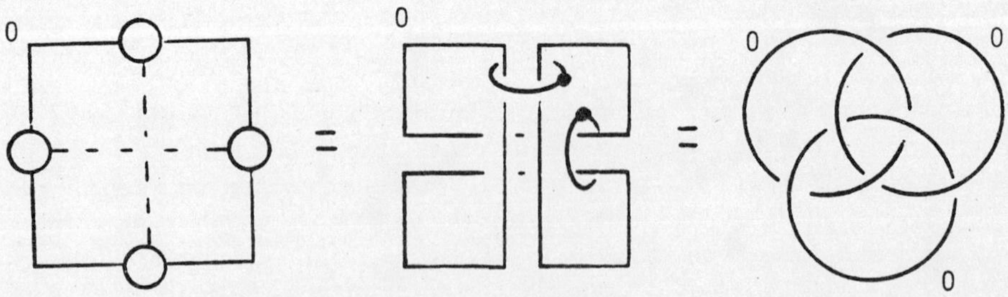

Figure 5.2

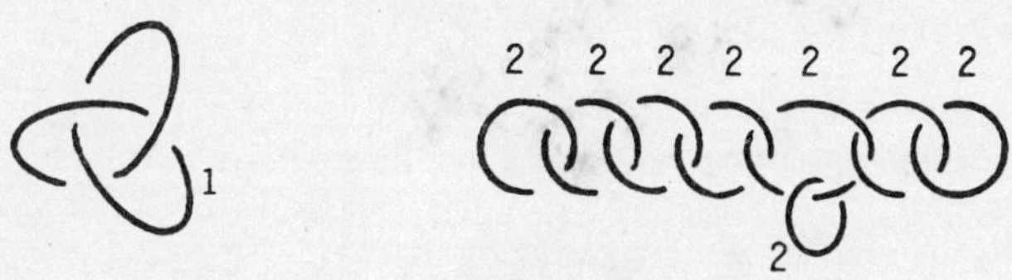

Figure 5.3

Example 5.4: Both links in Figure 5.3 describe the Poincaré homology 3-sphere.

Proof: The trefoil knot with framing 1 can be used as the definition of the Poincaré homology sphere (for others see [**K-Sh**]). To see that the second link has the same boundary, introduce three unknots with framing -1 and slide the endmost circles over them (Figure 5.4a). Now apply Lemma 4.1 to the three circles with framing 1 to get Figure 5.4b.

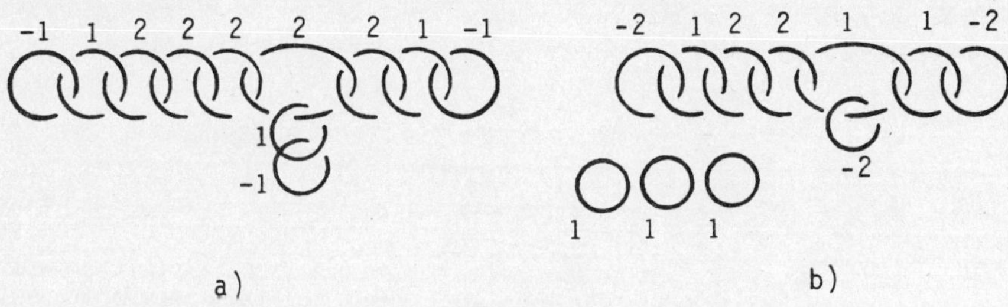

Figure 5.4

Iterate this process, discarding unknotted, unlinked components, to obtain Figure 5.5a. Apply Lemma 4.1 to a succession of unknots with -1 framing to finish the argument. □

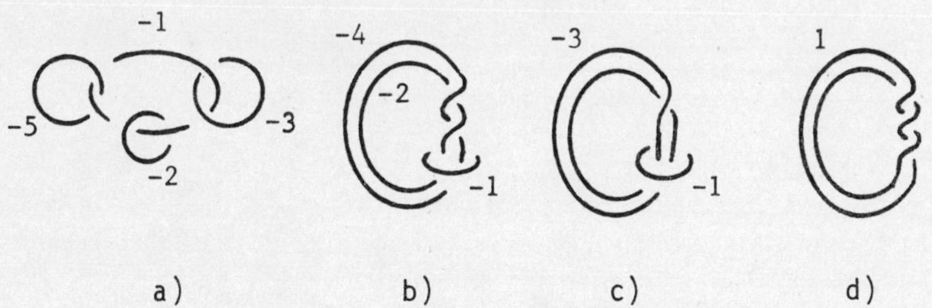

Figure 5.5

Theorem 5.1 does not seem to be useful for classifying 3-manifolds, for there has been little progress on any of the following questions: Given an arbitrary oriented 3-manifold N^3,

a) Find a "canonical" or "minimal" link L so that $\partial M_L = N^3$.

b) If $N^3 = \partial M_{L_1} = \partial M_{L_2}$ find an algorithm for moving from L_1 to L_2 using moves (1) and (2).
c) Solve (b) at the level of linking matrices using the algebraic analogue of moves (1) and (2).
d) Solve any of (a), (b) or (c) for interesting classes of 3-manifolds. (For Brieskorn homology 3-spheres, and more generally Seifert manifolds, there are interesting resolutions or graphs describing them; these have framed link analogues. See, for example, Walter Neumann's papers or [**N-R**].)

Our occasional use of moves (1) and (2) in Theorem 5.1 will be to check that ∂M_L is either S^3 or $\overset{k}{\#} S^1 \times S^2$ so as to be able to construct a closed 4-manifold.

§6. A Homotopy 4-Sphere.

As a final illustration of handlebody and framed link techniques, we construct an interesting homotopy 4-sphere [**A-K2**] [**A-K3**].

The framed link L in Figure 6.1 describes a contractible 4-manifold $\sum_0^4 = M_L$ ($\pi_1 = 0$ is shown below). To see that $\partial M_L = S^3$, surger the dotted circles to 2-handles with 0-framing (Lemma 2.1), shrink the circles with ± 1 framings so that they are small unknots and apply Lemma 4.1, and untangle the mess to obtain the unlink of four components with ± 1 framings, i.e. S^3.

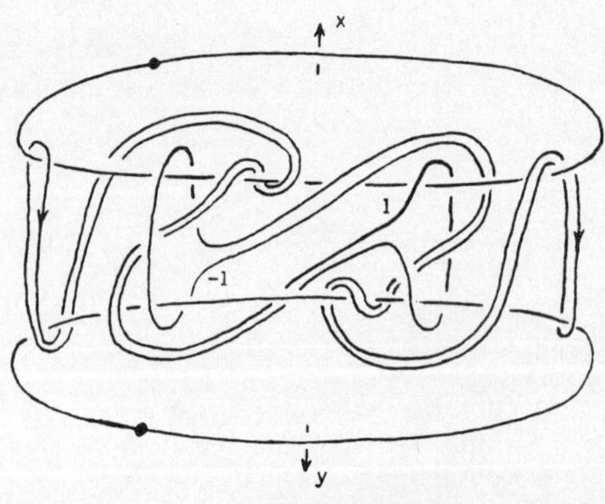

Figure 6.1

Thus $\sum^4 = \sum_0^4 \cup$4-handle is a homotopy 4-sphere. It was not created out of thin air, but is the double cover of an exotic smooth structure on RP^4 ([**C-S**] and [**Freedman2**]).

It is easy to construct "bushel baskets" of homotopy 4-spheres via the "Gluck construction" [**Gluck**]. Given a smoothly knotted 2-sphere Θ in S^4, remove its tubular neighborhood $S^2 \times B^2$ and glue it back in by the only interesting diffeomorphism of $S^1 \times S^2$, namely the one derived from the non-trivial element in $\pi_1(SO(3)) = Z/2$. The result $Q^4(\Theta)$ is easily seen to be a homotopy 4-sphere, but is not known to be diffeomorphic to S^4 except in a few cases. I know of no reason to guess whether an arbitrary $Q^4(\Theta)$ is S^4 or not, but because the M_L of Figure 6.1 comes from a "fake RP^4" it might also be a fake. In fact, M_L can also be described as $Q^4(\Theta)$ (see [**A-K3**],

Figure 16) where Θ is constructed from two different ribbon disks for the 8_9 knot drawn in Figure 6.2 (the two different ribbon moves are indicated by dotted lines). Notice the symmetry (rotation by π and reflection in plane of the paper) which takes one ribbon move to the other.

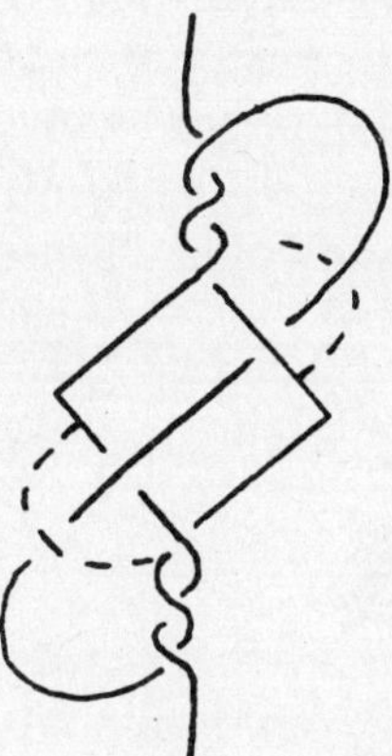

Figure 6.2

$\sum^4$ is homeomorphic to S^4 by Freedman's Theorem (III §1 and [**Freedman1**]), but more is easily shown. If we add two 2-handles to $\sum_0^4 = M_L$ as shown in Figure 6.3, then it is not hard to show, sliding the old 2-handles over the new ones in the obvious way, that the new manifold is diffeomorphic to $S^2 \times B^2 \, \natural \, S^2 \times B^2$. Then we can add two 3-handles and a 4-handle to get S^4.

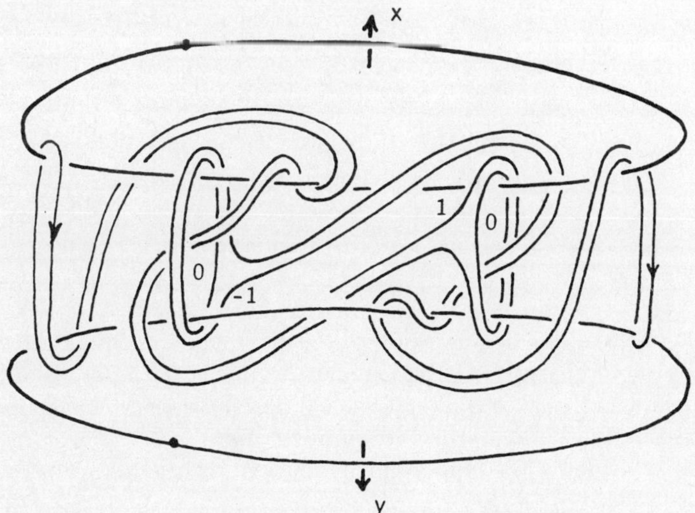

Figure 6.3 (Figure 29 of [**AK1**])

Thus $\sum_0 \subset S^4$ and $S^3 = \partial \sum_0$ is a smoothly imbedded 3-sphere in S^4. By the topological Schoenflies theorem ([**Mazur**], [**Brown**]), $\sum_0$ is then homeomorphic to B^4.

This raises the question of the piecewise linear (PL) and smooth Schoenflies Conjectures: A smooth (PL) imbedding of S^{n-1} in S^n bounds two smooth (PL) n-balls. The PL version is true in dimensions other than 4. The smooth version fails in higher dimensions because of exotic smooth structures on spheres. But in dimension 4, the PL and smooth categories are essentially equivalent [**Cerf2**]. So the smooth 4-dimensional Schoenflies conjecture (does every smooth 3-sphere in S bound a smooth 4-ball?) is wide open.

Scharlemann [**Sch**] has shown that it is true in the genus two case where S^3 has genus n if there is a Morse function on S^4 with two critical points which restricts to a Morse function on S^3 with k o-handles and $n+k-1$ 1-handles. In the above example, $S^3 = \partial \sum_0$ can be arranged to have genus 4 (W. Eaton and R. Gompf).

There is one more old conjecture related to $\sum^4$, the Andrews–Curtis conjecture. It is a purely group theoretic conjecture, but we motivate it as follows. From Figure 6.1, we can read off a presentation of the fundamental group of $\sum_0$; starting at the arrows on the 2-handles we get generators x and y and relations

$$1 = yx^{-2}yxy^{-2}x(xy^{-1}x^{-1}) \tag{1}$$

and

$$1 = yx^{-2}yxy^{-2}x(y^{-1}x^{-1}y) \tag{2}$$

so

$$xy^{-1}x^{-1} = y^{-1}x^{-1}y \tag{3}$$

or

$$xyx = yxy. \tag{4}$$

Using (3) and (4), we obtain from (1),

$$\begin{aligned} 1 &= yx^{-2}(yx)y^{-2}x^2(y^{-1}x^{-1}) \\ &= y(yx)y^{-4}(y^{-1}x^{-1})y^2 \\ &= x^{-5}(yx)x^{-1}y^3 \\ &= x^{-5}y^4. \end{aligned} \tag{5}$$

This presentation $\{x, y \mid xyx = yxy, x^5 = y^4\}$ is interesting. It is the trivial group since (4) implies that $y = (yx)^{-1}x(yx)$, so $y^5 = (yx)^{-1}x^5(yx) = (yx)^{-1}y^4(yx) = x^{-1}y^4x = x^{-1}x^5x = x^5 = y^4$ so $y = 1$ and $x = 1$.

The reduction of the original presentation to the trivial one may suggest how to slide handles in Figure 6.1 to change M_L to the 4-ball. Indeed, the steps that obtained identities (3), (4) and (5) can be carried out by handle slides. Since it is not clear which bands to use in the band connected sums of the handle slides, we can up the dimension by 1 and consider $\sum_0 \times I$ where attaching circles are unknotted and unlinked in $S^4 = \partial(B^5) = \partial(0\text{-handle})$.

However, the reduction of $\{x, y \mid xyx = yxy, x^5 = y^4\}$ cannot be carried out because we need extra 2-handles to "remember" relations. If relation r_1 is substituted in r_2, then handle r_2 is slid over handle r_1 and no longer exists as the handle r_2. But of course the relation r_2 is still valid and may be used later.

So the Andrews–Curtis Conjecture ([**A-C**], [**Kirby2**] Problem 5.2) states that a presentation of the trivial group can be moved to the trivial presentation by the usual Tietze moves except that one is not allowed to remember relations.

The presentations $\{x, y \mid xyx = yxy, x^{n+1} = y^n\}$ may be the simplest class not known to satisfy the conjecture. $\{a, b \mid a^{-1}b^2a = b^3, b^{-1}a^2b = a^3\}$ is another.

If a homotopy 4-ball B (with $\partial B = S^3$) is described as a handlebody with no 3-handles, and its presentation of the trivial group satisfies the Conjecture, then handle slides can be carried out in $B \times I$ to show that $B \times I = B^5$. Then ∂B is a smooth 3-sphere in $\partial B^5 = S^4$, so we are back to the smooth Schoenflies Conjecture; at least B is homeomorphic to B^4.

One can "remember" a 2-handle at the cost of adding a cancelling 2-3 handle pair. Then the 2-handle is slid over the 2-handle which is to be remembered, making a second copy to be saved. If this is done on the homotopy 4-ball itself there are usually geometric problems. If it is done on $B \times I$ to avoid unwanted linking, then the original 1- and 2-handles will be cancelled (following the algebra) but there will be 2- and 3-handles left which may not cancel; the 2-handles are added to trivial circles, but the 3-handles are complicated.

The questions and conjectures raised in this section are important but hard. There is some progress, e.g. [**Quinn2**], but not much.

(Added February 1988; Gompf has shown that $\sum_0$ is diffeomorphic to B^4; the argument involves handle slides and begins with a clever introduction of a 2-3 handle pair where the 2-handle is attached to a non-trivial circle which is in fact trivial in $\partial \sum_0$. The smooth Schoenflies conjecture is then settled in this particular case.)

II. INTERSECTION FORMS

§1. Intersection Forms.

Let M^4 be closed and oriented. If $\pi_1(M) = 0$, then $H_2(M;Z)$ and $H^2(M;Z)$ are free Z-modules of rank equal to the second Betti number. If $\pi_1(M) \neq 0$ then $H_2(M;Z)/T$ and $H^2(M;Z)/T$ are free Z-modules where T is the torsion subgroup of the appropriate group. There are two isomorphisms $H_2(M;Z)/T \xrightarrow{\cong} H^2(M;Z)/T$, one given by Poincaré duality and denoted by " ˆ ", as in $\alpha \to \hat{\alpha}$, and a second given by Hom and denoted by " * ", as in $\alpha \to \alpha^*$.

THEOREM 1.1. *Let M be closed, oriented and smooth. Any element $\alpha \in H_2(M;Z)$ is represented by a smoothly imbedded, oriented surface F_α. Another such surface F'_α representing α is joined by a smoothly imbedded oriented 3-manifold Y^3 in $M^4 \times I$ with $\partial Y = F'_\alpha - F_\alpha$.*

(Note that if $M = M_L$ for a framed link L with no 1-handles, then a basis can be represented by surfaces made up of cores of 2-handles union Seifert surfaces for their attaching circles.)

PROOF: There is an isomorphism

$$H^2(M;Z) \cong [M, K(Z,2)] = [M, CP^\infty]$$

where brackets denote homotopy classes of maps. So $\hat{\alpha} \in H^2(M;Z)$ corresponds to a homotopy class of maps $\{f\} : M \to CP^\infty$. Since $\dim M = 4$, $[M, CP^\infty] = [M, CP^2]$. Make f smoothly transverse to CP^1; $f^{-1}(CP^1) = F^2_\alpha$ will be an oriented surface representing α. F'_α will define a map $f' : M \to CP^2$ with $(f')^{-1}(CP^1) = F'_\alpha$ and the homotopy $f_t : M \times I \to CP^2$ between f and f', made transverse to CP^1, will provide Y^3. □

Note that the normal bundle for F_α or Y^3 is the pull back under f or f_t of the normal bundle of CP^1 in CP^2. The latter has Euler class 1 times the generator of $H^2(CP^1;Z) = H^2(S^2;Z) = Z$, so if the Euler class of the normal bundle to F_α, ν_{F_α} is n (times a generator), it follows that the map $f : F_\alpha \to CP^1$ must have degree n since $\nu_{F_\alpha} = f^*(\nu_{CP^1})$ and $f^*(\chi(\nu_{CP^1})) = \chi(\nu_{F_\alpha})$.

Also note there is a natural version of this theorem for manifolds with boundary, properly imbedded surfaces, and relative classes.

Furthermore, this theorem works for topological manifolds, but the full work of Freedman and Quinn is needed to establish the necessary transversality theorems [**Quinn1**].

In the case with $Z/2$-coefficients, a 2-dimensional homology class can be represented by a smoothly imbedded, unoriented 2-manifold, but two such cannot always be joined by a smooth, unoriented 3-manifold, even only immersed ([**F-K**], Theorem 3).

There is a pairing

$$H^2(M;Z)/T \otimes H^2(M;Z)/T \to Z$$

given by $\hat{\alpha}\otimes\hat{\beta} \to (\hat{\alpha}\cup\hat{\beta})[M] \in Z$ where $[M]$ is the fundamental class of M in $H_4(M;Z)$. Since in this dimension, $\hat{\alpha}\cup\hat{\beta} = \hat{\beta}\cup\hat{\alpha}$, the pairing is symmetric. Note that this pairing is defined on torsion elements, but is zero. It defines an integral, symmetric, bilinear form on $H^2(M;Z)/T$.

In the smooth case we can understand this pairing as the intersection form on $H_2(M;Z)/T$. Using Theorem 1.1, represent the Poincaré duals α and β to $\hat{\alpha}$ and $\hat{\beta}$ by smoothly imbedded, oriented surfaces F_α and F_β. Perturb either one so that F_α meets F_β transversely in n points, $p_1,\ldots,p_n$. Each point p_i can be assigned a + or -1, $\varepsilon(p_i) = \pm 1$, according to whether $T_{F_\alpha}|p_i \oplus T_{F_\beta}|p_i$ has the same or opposite orientation as $T_M|p_i$. Then define the intersection $\alpha\cdot\beta$ by $\alpha\cdot\beta = \sum_{i=1}^{n}\varepsilon(p_i)$, the algebraic number of points of intersection. This can be seen to be independent of the various choices by a bordism argument involving Y_α and F_α and F'_α. By Poincaré duality, $a\cdot\beta = (\hat{\alpha}\cup\hat{\beta})[M]$.

(Long ago I heard the dictum, "Think with intersections, prove with cup products." Being geometrically minded, I will go reasonably far (perhaps beyond?) with geometric arguments, but occasionally homotopy theoretic proofs are appropriate, as with Theorem 2.1 below.)

Given a basis $\alpha_1,\ldots,\alpha_k$ for $H_2(M;Z)$, then the intersection form determines the intersection matrix $(\alpha_i\cdot\alpha_j)$.

If $M^4 = M_L$, where L has only 2-handle components $L_1,\ldots,L_k$, then the intersection form with basis $\alpha_1,\ldots,\alpha_k$ determined by $L_1,\ldots,L_k$ is represented by the linking matrix for L. If one 2-handle is slid over another, say "L_i" over "L_j", then the homology basis changes by sending α_i to $\alpha_i+\alpha_j$ and the matrix changes by adding the jth row to the ith row and then the jth column to the ith column.

The intersection pairing is unimodular, that is, given a linear map $\lambda : H_2(M;Z) \to Z$, there exists an element $\alpha \in H_2(M;Z)$ such that $\lambda(\beta) = \alpha\cdot\beta$ for all β (see §2 below). For λ defines an element $\hat{\alpha} \in H^2(M;Z)$ and its Poincaré dual α satisfies $\lambda(\beta) = \alpha\cdot\beta$. An equivalent definition is that the intersection matrix has determinant ± 1.

If $\partial M \neq \emptyset$, then the intersection form is still well defined on $H_2(M;Z)$. In fact, if ∂M is a homology 3-sphere, then the above discussion goes through unchanged, since $M \cup (\text{cone } \partial M)$ looks to homology like a closed manifold. But for general ∂M, the intersection form is no longer unimodular; for example, let ξ_k be the B^2-bundle over S^2 with $\chi(\xi_k) = k \neq \pm 1$, and then the intersection form is represented by the 1×1 matrix (k).

Note that if $\alpha = i(\alpha')$, $i : H_2(\partial M;Z) \to H_2(M;Z)$ then $\alpha\cdot\beta = 0$ for all β. Then the intersection form is non-singular on $H_2(M;Z)/(T + i(H_2(\partial M;Z)))$ and is unimodular with field coefficients.

EXAMPLES: $H_2(S^2\times S^2;Z) = Z\oplus Z$ with generators $\alpha = S^2\times p$ and $\beta = q\times S^2$, so with appropriate orientations $\alpha\cdot\beta = \varepsilon(p,q) = 1$ and $\alpha\cdot\alpha = \varepsilon(S^2\times p \cap S\times p') = \varepsilon(\emptyset) = 0$ so the form is $\begin{pmatrix} 0 & 1 \\ 1 & 0 \end{pmatrix}$.

$H_2(CP^2; Z) = Z$ with generator $CP^1 = S^2$. As we have seen, the normal bundle of CP^1 is the Hopf bundle with first Chern class $c_1 = 1 =$ Euler class, so $S^2 \cdot S^2 = 1$ and the form is just (1). The form for CP^2 with the opposite orientation, $-CP^2$, is (-1). (Note that CP^2 has a preferred orientation, as do all complex manifolds.)

$H_2(M^4 \# N^4; Z) = H_2(M) \# H_2(N)$ and the form decomposes as a direct sum $\left(\begin{array}{c|c} \text{form}_M & 0 \\ \hline 0 & \text{form}_N \end{array}\right)$ so the form for $\overset{p}{\#} CP^2 \overset{q}{\#} (-CP^2)$ is

$$\begin{pmatrix} 1 & & & & & \\ & \ddots & & & 0 & \\ & & 1 & & & \\ & & & -1 & & \\ & 0 & & & \ddots & \\ & & & & & -1 \\ \multicolumn{3}{c}{\underbrace{\hphantom{1 \quad \ddots \quad 1}}_{p}} & \multicolumn{3}{c}{\underbrace{\hphantom{-1 \quad \ddots \quad -1}}_{q}} \end{pmatrix}$$

The framed link in Figure I.5.3 has $\partial M_L =$ Poincaré homology sphere, so its intersection matrix has determinant 1 and is

$$\begin{pmatrix} +2 & 1 & & & & & & \\ 1 & +2 & 1 & & & & & \\ & 1 & +2 & 1 & & & & \\ & & 1 & +2 & 1 & & & \\ & & & 1 & +2 & 1 & 0 & 1 \\ & & & & 1 & +2 & 1 & 0 \\ & & & & 0 & 1 & +2 & 0 \\ & & & & 1 & 0 & 0 & +2 \end{pmatrix}$$

§2. Homotopy Type.

THEOREM 2.1 [**Wh**], [**Milnor1**], [**M-H**]. *If M and M' are simply connected, closed, oriented 4-manifolds, then they are homotopy equivalent iff their intersection forms are isomorphic.*

Later, Lemma IX.4, we show that the homotopy equivalence is a tangential homotopy equivalence, i.e. it is covered by a bundle map from T_M to T'_M.

PROOF: The proof on pages 104–105 of [**M-H**] is short and readable; we follow the outline here but make the end more geometric. Let M_0 be M minus the interior of a 4-ball. Then $H_i(M_0; Z) = 0$ except for $H_0(M_0; Z) = Z$ and $H_2(M_0; Z) = Z^r$. Since $\pi_1(M) = \pi_1(M_0) = 0$, it follows that $\pi_2(M_0) \cong H_2(M_0; Z) \cong Z^r$, so there exists a map $f : S^2 \vee \overset{r}{\ldots} \vee S^2 \to M_0$ which induces a homology isomorphism $f_* : H_*(S^2 \vee \overset{r}{\ldots} \vee S^2; Z) \xrightarrow{\cong} H_*(M_0; Z)$. If M_0 is a CW-complex, then f is a homotopy equivalence; in the smooth case, M_0 is a CW-complex, and in the topological case, M_0 is an absolute neighborhood retract and hence the homotopy type of a CW-complex which suffices.

Now M is obtained from M_0 by attaching a 4-cell, so M has the homotopy type of a space obtained from the r-fold wedge $S^2 \vee \cdots \vee S^2$ by attaching a 4-cell by an attaching map $g : S^3 \to S^2 \vee .\overset{r}{.}. \vee S^2$. So the homotopy type of M is determined by the homotopy class of g in $\pi_3(S^2 \vee .\overset{r}{.}. \vee S^2)$.

We show that an element of $\pi_3(S^2 \vee .\overset{r}{.}. \vee S^2)$ is determined by a certain linking matrix which coincides with the intersection form for M. Let $p_i \in S_i^2$ be a point (other than the base point) on the ith 2-sphere of $S^2 \vee \cdots \vee S^2$. Make g transverse to p_i, $i = 1, \ldots, r$, and let $L_i = g^{-1}(p_i)$. Then each L_i is an oriented (using the orientations of M^4, S^3, S^2) framed link with framing on a component L_{ij} equal to the difference between the 0-framing and the framing pulled back by g from a normal disk to p_i in S_i^2. L_i then has a linking matrix $\{a_{jk}\}$ as in I, §2. If we homotope g so that $g^{-1}(p_i)$ has just one component, i.e. if we band connect sum the components of L_i, then the framing of the one component, λ_i, is $\sum\limits_{j,k} a_{jk} =$ sum of the entries of $\{a_{jk}\}$.

Now $L = \bigcup\limits_{i=1}^{r} L_i = \bigcup\limits_{i=1}^{r} \lambda_i$ is itself a framed link with a linking matrix A. The proof is finished with the following two assertions:

Assertion 1: A is isomorphic to the intersection form on $H_2(M; Z)$.

Assertion 2: There is a one-to-one correspondence between symmetric $r \times r$ matrices A and elements of $\pi_3(S^2 \vee .\overset{r}{.}. \vee S^2)$.

PROOF OF ASSERTION 1: If we let each λ_i bound a surface F_i in the 4-cell B^4, then $\overline{F}_i$ in $S^2 \vee \cdots \vee S^2 \cup_g B^4$, defined by $F_i/\partial F_i$ where $g(\partial F_i) = g(\lambda_i) = p_i$ is a closed surface. The $\overline{F}_i$, $i = 1, \ldots, r$, generate $H_2(S^2 \vee \cdots \vee S^2 \cup_g B^4; Z) = H_2(M; Z)$ and clearly the intersection matrix of $\overline{F}_i$ both represents the intersection form on $H_2(M; Z)$ and is A.

PROOF OF ASSERTION 2: We have already constructed a map from $g : S^3 \to S^2 \vee \cdots \vee S^2$ to a matrix A via the Thom–Pontragin construction; the same methods show that homotopic maps give framed bordant links and hence the same matrix A. It is easy to construct a framed link (hence a map g) which gives a prescribed matrix A. We leave to the reader the task of showing that if two framed links have the same linking matrix, then they are framed bordant.

This proof is more round-about than that in **[M-H]**, but it makes the geometric connection between homotopy type and intersection forms clear.

§3. Symmetric Bilinear Forms.

In this section we collect some examples and results about integral, symmetric, unimodular bilinear forms.

We have already met the forms whose matrices (in terms of a suitable basis) are (1), (-1), $\begin{pmatrix} 0 & 1 \\ 1 & 0 \end{pmatrix}$ and their direct sums. Another example is Γ^{4k} which is the lattice in R^{4k} generated by $e_i + e_j$ and $1/2(e_1 + e_2 + \cdots + e_{4k})$ where $e_1 \ldots e_{4k}$ is the usual orthonormal basis for R^{4k} and the form is the restriction of the usual Euclidean inner product on R^4. Thus $(e_i + e_j)^2 = 2$ and $(1/2(e_1 + \cdots + e_{4k}))^2 = k$. Γ^{4k} is positive definite.

For $k = 2$, Γ^8 is isomorphic to a well known form which is often called E_8 (since it arises from the Dynkin diagram for the Lie group E_8); one choice of matrix for E_8 is

$$\begin{pmatrix}
2 & 1 & & & & & & \\
1 & 2 & 1 & & & & & \\
 & 1 & 2 & 1 & & & & \\
 & & 1 & 2 & 1 & & & \\
 & & & 1 & 2 & 1 & 0 & 1 \\
 & & & & 1 & 2 & 1 & 0 \\
 & & & & 0 & 1 & 2 & 0 \\
 & & & & 1 & 0 & 0 & 1
\end{pmatrix}$$

which comes from the weighted tree (compare the last example in §1),

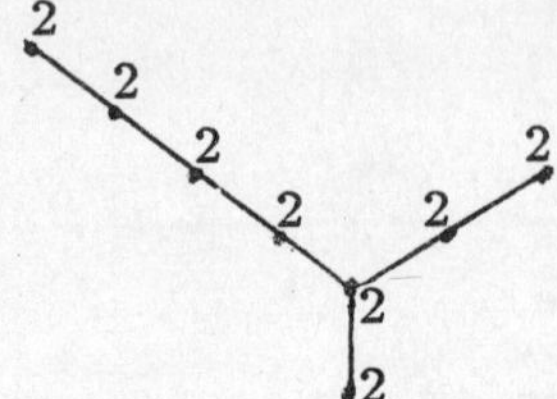

One more example is the form L arising from the Leech lattice. In R^{25} choose the metric $(+,+,+,\ldots,+,-)$, that is, $(x_1,\ldots,x_{25})\cdot(y_1,\ldots,y_{25}) = \sum_{i=1}^{24} x_i y_i - x_{25}y_{25}$. Let L be the orthogonal complement of the vector $\lambda = (3,5,7,9,\ldots,43,45,47,51)$. Since $\lambda\cdot\lambda = -1$, $L = \lambda^\perp$ is positive definite. Furthermore, $x\cdot x \geq 4$ and $x\cdot x \equiv 0(2)$ for all x.

Formally, an integral bilinear form consists of a free Z-module X and a bilinear form ϕ,

$$X \otimes X \xrightarrow{\phi} Z.$$

It is symmetric if $\phi(x_1,x_2) = \phi(x_2,x_1)$. It is unimodular if for a basis $x_1,\ldots,x_k$, $\det(\phi(x_i,x_j)) = \pm 1$, or equivalently, if $\lambda : X \to Z$ is linear, then there exists a unique $y_0 \in X$ such that $\lambda(x) = \phi(x,y_0) = \phi(y_0,x)$, or equivalently, given a basis $x_1,\ldots,x_k$, there exists a dual basis $y_1,\ldots,y_k$ with $\phi(x_i,y_j) = \delta_{ij}$. It is positive (negative) definite if $x\cdot x > 0(< 0)$ for all $x \in X$. It is indefinite otherwise. Often we write $\phi(x,y) = x\cdot y$ and think of X as $H_2(M;Z)$.

We assume throughout these notes that ϕ is symmetric, unimodular and integral unless otherwise stated.

There are three invariants of ϕ:

1) rank ϕ = dimension of X as a free Z-module.
2) type ϕ, which is *even* if $x\cdot x \equiv 0(2)$ for all $x \in X$, and *odd* if $\exists x$ such that $x\cdot x \equiv 1(2)$.
3) signature ϕ, which is the number of positive entries minus the number of negative entries if we diagonalize over the rationals. (Note that this definition makes sense without unimodularity, i.e. in the case of $H_2(M;Z)$ when $\partial M \neq 0$.) In the indefinite case, these invariants determine the form ([M-H], Theorem II.5.3).

First we quote a lemma of Meyer ([M-H], p. 22).

LEMMA 3.1. *If ϕ is indefinite, then there exists $x \in X$ such that $x \cdot x = 0$. Clearly x can be chosen indivisible, i.e. not a multiple of another element.*

Then in the odd case we have:

THEOREM 3.2. *If ϕ is odd and indefinite, then ϕ decomposes as a direct sum $\phi = \overset{p}{\oplus}(1) \overset{q}{\oplus}(-1)$.*

PROOF: Choose an indivisible x with $x \cdot x = 0$ and y' with $x \cdot y' = 1$ and let Y be the orthogonal complement of the subspace generated by x and y'. If $y' \cdot y'$ is even, then Y is odd, so we can choose an odd element $y'' \in Y$ and then $(y' + y'')^2$ is odd. So we can choose x and y with matrix $\begin{matrix} x \\ y \end{matrix}\begin{pmatrix} 0 & 1 \\ 1 & \text{odd} \end{pmatrix}$. For the correct choice of $k \in Z$, we have $\begin{matrix} x \\ y - kx \end{matrix}\begin{pmatrix} 0 & 1 \\ 1 & 1 \end{pmatrix}$. Then $\begin{matrix} x - (y - kx) \\ y - kx \end{matrix}\begin{pmatrix} -1 & 0 \\ 0 & 1 \end{pmatrix}$. Choose either $x - (y - kx)$ or $y - kx$ so that its orthogonal complement is indefinite, and repeat the argument until ϕ is decomposed as desired. □

DEFINITION: Call $\omega \in X$ characteristic if $\omega \cdot x \equiv x \cdot x(2)$ for all $x \in X$.

(If $X = H_2(M^4; Z)$, then ω is an integral dual to the second Steifel–Whitney class of M^4, as we see later.)

LEMMA 3.3. *There exists a characteristic element ω with $\omega \cdot \omega$ well defined modulo 8.*

PROOF: Let $X_{(2)}$ be the mod 2 reduction of X and let $\overline{x} = x_{(2)}$. We have a homomorphism

$$\begin{aligned} X_{(2)} &\xrightarrow{h} Z/2 \\ \overline{x} &\longrightarrow \overline{x} \cdot \overline{x}. \end{aligned}$$

$\phi_{(2)}$ is an inner product on the vector space $X_{(2)}$, so the linear map h is given by inner product with a fixed element $\overline{\omega} \in X_{(2)}$, i.e. $\overline{x} \cdot \overline{x} = \overline{\omega} \cdot \overline{x}$. Then let ω be any element in X which reduces mod 2 to $\overline{\omega}$. Clearly $\omega \cdot x \equiv x \cdot x(2)$. Furthermore, if ω' is another such, then $\omega' = \omega + 2x$, so $\omega' \cdot \omega' = (\omega + 2x) \cdot (\omega + 2x) = \omega \cdot \omega + 4\omega \cdot x + 4x \cdot x = \omega \cdot \omega + 4(\text{even}) = \omega \cdot \omega \bmod(8)$. □

LEMMA 3.4. *signature $\phi \equiv \omega \cdot \omega$ (8).*

PROOF: Since every odd, indefinite form ϕ is isomorphic to $\overset{p}{\oplus}(1) \overset{q}{\oplus}(-1)$, sign $\phi = p - q = \omega \cdot \omega$ since we can choose ω to be the sum of the generators of the factors. In general, consider $X \oplus (1) \oplus (-1)$ which is odd and indefinite. Then

$$\text{sign}\, X = \text{sign}(X \oplus (1) \oplus (-1)) \overset{(8)}{\equiv} (\omega_x + \alpha + \beta) \cdot (\omega_x + \alpha + \beta) = \omega_x \cdot \omega_x$$

where ω_x is characteristic for X and α and β generate (1) and (−1). □

In the even case we can choose $\omega = 0$ and thus sign $\phi \equiv 0(8)$. Furthermore, rank ϕ and sign ϕ are even, since as long as ϕ is indefinite, we can split off direct summands

$\begin{pmatrix} 0 & 1 \\ 1 & 0 \end{pmatrix}$; the remaining definite form has rank = index = even. We can then represent all possible ranks and signatures by the forms

$$\pm \overset{r}{\oplus} E_8 \overset{s}{\oplus} \begin{pmatrix} 0 & 1 \\ 1 & 0 \end{pmatrix}.$$

It is a fact ([M-H], Theorem II.5.3) that in the indefinite case two forms are isomorphic if they have the same rank, type, and signature, so these are the only forms.

We summarize our information in a table

<table>
<tr><th></th><th>indefinite</th><th>definite</th></tr>
<tr><td>odd</td><td>$\overset{p}{\oplus}(1)\overset{q}{\oplus}(-1) = \begin{pmatrix} 1 & & & & & \\ & \ddots & & & & \\ & & 1 & & & \\ & & & -1 & & \\ & & & & \ddots & \\ & & & & & -1 \end{pmatrix}$
$p+q =$ rank
$p-q =$ signature
represented by $\overset{p}{\#}\, CP^2 \overset{q}{\#}\, (-CP^2)$</td><td>$\pm\overset{p}{\oplus}(1)$, represented by $\overset{p}{\#} \pm CP^2$,
and many other forms, e.g.
$E_8 \oplus (1)$ and $\Gamma^{4(2k+1)}$</td></tr>
<tr><td>even</td><td>$\pm \overset{r}{\oplus} E_8 \overset{s>0}{\oplus} \begin{pmatrix} 0 & 1 \\ 1 & 0 \end{pmatrix}$
$8r + 2s =$ rank
$\pm 8r =$ signature
$\begin{pmatrix} 0 & 1 \\ 1 & 0 \end{pmatrix}$ is represented by $S^2 \times S^2$
$E_8 \oplus E_8 \overset{3}{\oplus} \begin{pmatrix} 0 & 1 \\ 1 & 0 \end{pmatrix}$ is represented by the Kummer surface.</td><td><table>
<tr><td>rank</td><td>8</td><td>16</td><td>24</td><td>32</td><td>40</td></tr>
<tr><td># of forms</td><td>1</td><td>2</td><td>24</td><td>$> 10^7$</td><td>$> 10^{51}$</td></tr>
<tr><td></td><td>$E_8 =$
Γ^8</td><td>Γ_{16}
$E_8 \oplus E_8$</td><td>Γ^{24}
L
⋮</td><td>Γ^{32}
.
⋮</td><td>Γ^{40}
.
⋮</td></tr>
</table></td></tr>
</table>

We have only listed examples of definite forms and some estimates of their numbers (only finitely many of given rank). Also we have indicated which odd forms are known to be represented by simply connected, smooth, closed 4-manifolds and the simplest even form represented by such.

Of course, any form can be represented by a 4-manifold with boundary. Just take any framed link L in S^3 whose linking matrix represents the form; then M_L^4 is a simply connected, smooth 4-manifold whose boundary is a homology 3-sphere if the form is unimodular. (Freedman has shown [**Freedman1**] that each homology 3-sphere bounds a contractible topological 4-manifold, so each form is represented by a simply connected, closed, topological 4-manifold.)

§4. Characteristic Classes.

The Stiefel–Whitney classes of T_M, $\omega_k(T_M) \in H^k(M^4; Z/2)$ are obstructions to finding a field of $4-k+1$ frames over the k-skeleton, ([**M-S**], §12). In the case $k=1$, ω_1 is the only obstruction and hence M is orientable iff $\omega_1 = 0$. Again, if $k=2$, ω_2 is the only obstruction to finding a field of 3-frames, hence 4-frames (using the orientation to choose a fourth vector over the 2-skeleton). A trivialization of T_M over the 2-skeleton is called a spin structure on M (see Chapter IV). Note that when $\omega_2 \neq 0$, T_M can still be trivialized on $M - F^2$ where F^2 is a surface dual to ω_2. ω_3 is always zero since $\pi_2(SO(4)) = 0$, so a spin 4-manifold has a trivialization of T_M over the 3-skeleton, hence on M^4-point. Such a manifold is called almost parallelizable.

ω_4 is not the only obstruction to extending the trivialization of T_M over the last point; both the first Pontrjagin class (or the index) and the Euler class of M must be zero (see Chapter VI).

We will usually work with oriented 4-manifolds, so $\omega_1 = 0$ and ω_2 will be the interesting characteristic class.

According to Wu's formula ([**M-H**], p. 136), on a closed, smooth, connected, oriented 4-manifold ω_2 is characterized by $\omega_2 \cup x = x \cup x$ for all $x \in H^2(M; Z/2)$.

As above, a characteristic element for the intersection form is $\omega \in H_2(M;Z)$ where $\omega \cdot x = x \cdot x(2)$ for all $x \in H_2(M;Z)$. When $H_1(M;Z) = 0$, then the relation between ω_2 and ω is simple: ω is an integral Poincaré dual to ω_2. That is, ω_2 is the mod 2 reduction of $\hat{\omega}$. For $H_1(M;Z) = 0$ implies that $H^2(M;Z) \to H^2(M;Z/2)$ is onto, so that ω_2 has an integral lift $\hat{\omega}$ and $\hat{\omega} \cup \hat{x} = \hat{x} \cup \hat{x}(2)$ for all $\hat{x} \in H^2(M;Z)$, thus $\omega \cdot x = x \cdot x(2)$. Thus:

LEMMA 4.1. *If $H_1(M;Z) = 0$ then $\omega_2 = 0$ iff the intersection form is even.*

However, when $H_1(M;Z) \neq 0$, $\omega_2 = 0$ implies that the intersection form is even, but the converse is not necessarily true. If ω_2 comes from an integral class $\hat{\omega}$, then its dual satisfies $\omega \cdot x = x \cdot x(2)$. But if ω_2 does not lift to an integral class, then it may happen that $\omega_2 \neq 0$ but the intersection form on $H_2(M;Z)$ is even.

An example from [**Habegger**] is $M = S^2 \times S^2/Z/2$ where $Z/2$ acts on $S^2 \times S^2$ by sending $(x,y) \to (-x,-y)$. The rank of $H^2(M;Z)$ is zero because rank $(H^2(M;Z)+2 = \chi(M) = 1/2\chi(S^2 \times S^2) = 2$. Furthermore, the diagonal S^2 in $S^2 \times S^2$ becomes an RP^2 in M with self-intersection 1, so its dual $x \in H^2(M;Z/2)$ forces $1 = x \cup x = \omega_2 \cup x$ so $\omega_2 \neq 0$.

A framed link description of M is the double DN_L of the 4-manifold N_L in Figure 4.1. (For doubles, see I, §2.) It is easy to check by surgering the 1-handle that the boundary of DN is $S^1 \times S^2$ so that a 3- and 4-handle can be added. The long 2-handle gives the RP^2 by letting its attaching circle bound a Mobius band which of course goes over the 1-handle. With the basis from the 2-handles, $H_2(M;Z/2) = Z/2 \oplus Z/2$ with intersection form $\begin{pmatrix} 1 & 0 \\ 0 & 0 \end{pmatrix}$ where $RP^2 \cdot RP^2 = 1$ and $\hat{\omega}_2 \cdot \hat{\omega}_2 = 0$. To take the double cover of M via framed links is an interesting exercise; for some help, consult [**A-K1**].

§5. The Index.

The *index* of an arbitrary oriented 4-manifold M^4, $\sigma(M)$ or index M, is the signature of its intersection form. Note that if we change the orientation of M, we change the sign

Figure 4.1

of index M, i.e. index$(-M) = $ $-$index M. Complex surfaces have preferred orientations coming from their complex structures ((x_1, y_1, x_2, y_2) gives the same orientation as (x_2, y_2, x_1, y_1) where (z_1, z_2) are local coordinates and $z_j = x_j + iy_j$) so index$(CP^2) = 1$ and index$(-CP^2) = -1$.

LEMMA 5.1. *Suppose an $r+s$ manifold V^{r+s} bounds an oriented $r+s+1$ manifold W. Assume rational coefficients and let $i_* : H_*(V) \mapsto H_*(W)$ be induced by the inclusion map.*

1) *If $x \in H_r(V)$ and $y \in H_s(V)$ and $i_*(x) = 0 = i_*(y)$ then $x \cdot y = 0$.*
2) *If $x \in H_r(V)$ and $i_*(x) = 0$, then for any $y \in H_s(V)$ with $x \cdot y \neq 0$ it follows that $i_*(y) \neq 0$.*
3) *If $x \in H_r(V)$ and $i_*(x) \neq 0$, then there exists a $y \in H_s(V)$ with $x \cdot y = 1$ and $i_*(y) = 0$.*

PROOF: 1) Since x and y are rational classes, multiples of them, px and qy, are integral; then px and qy are represented by oriented, transverse surfaces in V which bound oriented, transverse, 3-manifolds in W. These 3-manifolds intersect in arcs whose end points have opposite sign in V and therefore sum equal to zero, so $0 = px \cdot qy = x \cdot y$.

2) This is a equivalent to 1).

3) Since $0 \neq i_*(x) \in H_r(W)$, there is a non-zero dual y' to $i_*(x)$ in $H_{s+1}(W, \partial)$. Let $y = \partial(y')$ and observe that $x \cdot y = i_*(x) \cdot y' = 1$ and $0 = i_*(y) = i_*\partial(y')$ by exactness.

□

THEOREM 5.2. *If M^4 bounds an orientable 5-manifold W^5, then $\sigma(M) = 0$.*

PROOF: If $x \in H_2(M; Q)$ bounds in W, i.e. $i_* x = 0$, then choose a dual $y \in H_2(M; Q)$. The intersection form on the subspace $\{x, y\}$ spanned by x and y is $\begin{matrix} x \\ y \end{matrix}\begin{pmatrix} 0 & 1 \\ 1 & * \end{pmatrix}$ (since $x \cdot x = 0$ by 1) in Lemma 5.1) which has index equal to 0. Continue this process on the orthogonal complement to $\{x, y\}$ so as to split off another 2-dimensional subspace, and then another, and so on until $H_2(M; Q)$ is exhausted. Because of 3) in Lemma 5.1, if there are any elements left in the orthogonal complement at any stage, then there must be an element which bounds in W. Thus $H_2(M; Q)$ splits as a direct sum of 2-dimensional subspaces on which the intersection form has index zero; thus $\sigma(M) = 0$.

□

Clearly index$(M_1 \sharp M_2) = $ index $M_1 +$ index M_2. This equation still holds if M_1 and M_2 have the same boundary and we glue them together along ∂M_i.

THEOREM 5.3 (Novikov). *If* $\partial M_1 = -\partial M_2$ *as oriented manifolds, and* $M = (M_1 \cup M_2)/\partial M_1 \sim -\partial M_2$, *then* index $M =$ index $M_1 +$ index M_2.

First, here's the idea behind the proof. Let N^3 (unoriented) denote the manifolds ∂M_1, ∂M_2 and the submanifold of M equal to the identification of ∂M_1 with ∂M_2. Now let y be an element of $H_2(M)$ and intersect y with N. If the 1-cycle $y \cap N$ bounds in N, then y splits as a sum of elements in $H_2(M_1)$ and $H_2(M_2)$. But if $y \cap N$ does not bound in N then it represents a non-zero class in $H_1(N)$ which has a dual $x \in H_2(N;Q)$. The intersection form on the subspace generated by x and y is $\begin{matrix} x \\ y \end{matrix}\begin{pmatrix} 0 & 1 \\ 1 & * \end{pmatrix}$ with index zero ($x \cdot x = 0$ because x can be pushed off itself using a normal vector field to N in M). Here are the details

PROOF: We use rational coefficients throughout the proof. The Mayer–Vietoris sequence for $M = M_1 \cup M_2$ is

$$\longrightarrow H_2(N) \xrightarrow{(i_1,-i_2)} H_2(M_1) \oplus H_2(M_2) \xrightarrow{j_1+j_2} H_2(M) \xrightarrow{\partial} H_1(N) \longrightarrow .$$

Then $H_2(M) \cong (H_2(M_1) \oplus H_2(M_2))/\text{image}(i_1,-i_2) \oplus \text{image}\,\partial$. In the vector space $H_2(N)$, let η_1 and η_2 be the subspaces of elements which bound in M_1 and M_2 respectively; i.e. $\eta_i = \ker(H_2(\partial M_i) \to H_2(M_i))$, $i = 1,2$.

It is not hard to check that the following sequence is exact:

$$0 \longrightarrow \eta_1 + \eta_2 \longrightarrow H_2(N) \xrightarrow{(i_1,i_2)} H_2(M_1) \oplus H_2(M_2)/\text{image}(i_1,-i_2) \xrightarrow{\lambda}$$
$$(H_2(M_1)/H_2(\partial M_1)) \oplus (H_2(M_2)/H_2(\partial M_2)) \longrightarrow 0.$$

((1) if $z \in \eta_1$, then $(i_1(x), i_2(x)) = (0, i_2(x)) = (0, -i_2(-x)) = (i_1(-x), -i_2(-x))$ which is zero in $H_2(M_1) \oplus H_2(M_2)/\text{image}(i_1,-i_2)$; (2) if $(i_1(z), i_2(z)) = (i_1(w), -i_2(w))$ then $i_1(z-w) = 0 = i_2(z+w)$ so write $z = \frac{z-w}{2} + \frac{z+w}{2}$; (3) $\lambda(i_1(z), i_2(z))$ is obviously zero; (4) if $\lambda(x,y) = 0$, then x and y separately come from $H_2(N)$, so $\left(x + \frac{y-x}{2}, y - \frac{y-x}{2}\right) = \left(\frac{x+y}{2}, \frac{x+y}{2}\right)$ is in the image of (i_1, i_2).) Thus

$$H_2(M) \cong H_2(N)/(\eta_1 + \eta_2) \oplus H_2(M_1)/H_2(\partial M_1) \oplus H_2(M_2)/H_2(\partial M_2) \oplus \text{image}\,\partial.$$

$\eta_1 + \eta_2$ is precisely the kernel of $i_* : H_2(N) \to H_2(M)$, for if $x \in H_2(N)$ and $i_*(x) = 0$, then let c be a 3-chain in M which x bounds. We can choose c so that a neighborhood of x in c is a collar of x in either M_1 or M_2, and so that c is transverse to N. Then let $c_1 = \text{closure}(c \cap \text{int}\, M_1)$ and $c_2 = \text{closure}(c \cap \text{int}\, M_2)$ and note that $x = \partial c_1 + \partial c_2$ lies in $\eta_1 + \eta_2$.

Let $y = \partial \overline{y} \in$ image ∂ and let $x \in H_2(N)$ be dual to y; it follows that $i_*(x) \cdot \overline{y} = 1$, so $i_*(x)$ is dual to $\overline{y}$ in $H_2(M)$, so $i_*(x) \neq 0$ and x does not belong to $\eta_1 + \eta_2$. So image ∂ is dual to a subspace η_3 of $H_2(N)$ with $\eta_3 \cap \eta_1 + \eta_2 = 0$. Furthermore, $\eta_1 + \eta_2$ and η_3 span $H_2(N)$ since if $x \in H_2(N)$ does not belong to $\eta_1 + \eta_2$, then $i_*(x) \neq 0$, and $i_*(x)$ has a dual $\overline{y} \in H_2(M)$ and then x is dual to $\partial \overline{y}$. Thus $\dim \eta_3 = \dim(\text{image}\,\partial)$ and $H_2(N)/(\eta_1 + \eta_2) \cong \eta_3 \cong$ image ∂.

Now let $x_1, \ldots, x_k$ be a basis for η_3 and $y_1, \ldots, y_k$ a dual basis for image ∂ with $y_i = \partial(\overline{y}_i)$ so that $i_*(x_i) \cdot \overline{y}_j = \delta_{ij}$. Then the intersection form on $H_2(M)$ restricts to a form with index zero on a subspace isomorphic to $H_2(N)/(\eta_1+\eta_2)\oplus$ image ∂ with basis $i_*x_1,\ldots,i_*x_k,\overline{y}_1,\ldots,\overline{y}_k$, since the form is $\begin{matrix} i_*x_i \\ \overline{y}_i \end{matrix}\begin{pmatrix} 0 & 1 \\ 1 & * \end{pmatrix}$. Thus the index of M is the index of the form on $H_2(M_1)/H_2(\partial M_1)\oplus H_2(M_2)/H_2(\partial M_2)$ which index M_1+ index M_2. □

Novikov's additivity theorem easily fails if M_1 and M_2 are glued together along only part of their boundaries. The Hopf disk bundle M has boundary equal to S^3 and form equal to (1) and index $=1$, yet it is the union of the disk bundle over the two hemispheres, each of which is a 4-ball with index $=0$.

However, the additivity formula can be corrected when $M = M_1 \bigcup_N M_2$ and $N = \partial M_1 \cap \partial M_2$ is only part of ∂M_i. The idea is to form the manifold $M_0 = (\partial M_1 \times I \cup_{N\times 0} (\partial M_2 \times I)$ and compute its index, which depends on how $H_1(\partial N; Z)$ dies under the inclusion of ∂N in N, $\partial M_1 - N$, and $\partial M_2 - N$. (In particular, if $\partial N = S^2$ then index$(M_0)=0$.) Then

$$\text{index } M = \text{ index } M_0 + \text{ index } M_1 + \text{ index } M_2$$

as is easily seen from Figure 5.1 and the above Theorem 5.3. Details can be found in [**Wall5**].

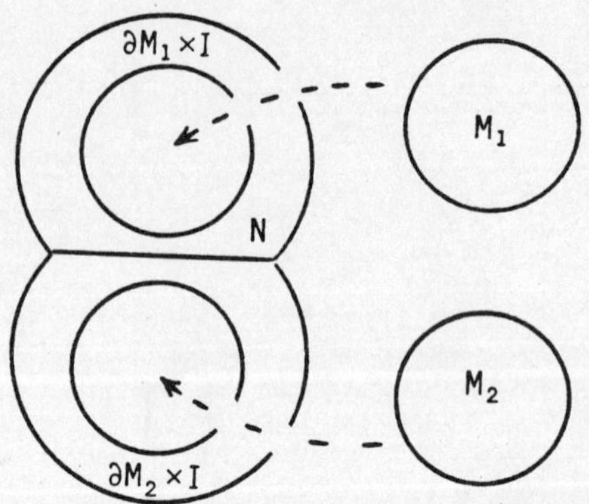

Figure 5.1

III. CLASSIFICATION THEOREMS

The classification problem for simply connected, closed 4-manifolds asks which homotopy types, that is, which forms (symmetric, integral, unimodular) can be represented by such manifolds and, if so, by how many.

§1. Rohlin's Theorem.

In the smooth case, the classical theorem is Rohlin's ([**Rohlin**], [**G-M**], [**F-K**]) which we prove in Chapter XI.

THEOREM 1.1. *If M^4 is closed, smooth and $\omega_1 = \omega_2 = 0$ (equivalently spin, or almost parallelizable, but even form is not sufficient [**Habegger**]), then* index $M \equiv 0(16)$.

Thus, "half" the even forms cannot be represented by a simply connected, smooth closed 4-manifold (see (2.6)). Rohlin's theorem can easily be generalized ([**K-M**], [**F-K**], Corollary XI.6) to

THEOREM 1.2. *If M^4 is closed, smooth, simply connected and if $\omega \in H_2(M; Z)$ is an integral dual to the second Stiefel–Whitney class ω_2, then ω can be represented by an imbedded 2-sphere which is smooth except at one point which is the cone on a knot K, and we have the congruence,*

$$\frac{\text{index } M - \omega \cdot \omega}{8} \equiv \text{Arf invariant}(K) \text{ modulo } 2.$$

The theorem does not eliminate any more forms, but does limit the 2-spheres representing ω.

§2. Freedman's Work.

In 1981 Freedman completely answered the classification problem in the topological case.

THEOREM 2.1 [**Freedman1**]. *Given an even (odd) form, there exists exactly one (two) simply connected, closed, topological 4-manifold(s) representing that form.*

In the odd case, the two manifolds are distinguished by their Kirby–Siebenmann triangulation invariants [**K-S**]. One has non-zero invariant and cannot be smooth (= piecewise linear in dim 4) and the other has zero invariant but still might not be smooth for other reasons, e.g. $E_8 \oplus 1$.

Freedman shows that every homology 3-sphere bounds a contractible 4-manifold. Thus, given a form ϕ, choose a framed link L with linking matrix ϕ, add 2-handles to B^4 to form M^4_L, and then cap off by adding Freedman's contractible 4-manifold to ∂M^4_L.

The form (1) is represented by ◯^{1} $= CP^2$ and also by (trefoil)^1 whose boundary is the Poincaré homology 3-sphere which is then capped.

It is worth noting that Freedman's theorem extends to the non-simply connected case when the fundamental group does not grow too fast; in particular it holds [**Freedman2**] for the class of fundamental groups which

1) contains Z and finite groups, and
2) is closed under the operations: subgroup, quotient, extension, and (infinite) nested union.

§3. Donaldson's Work.

A few months after Freedman's surprise, Donaldson proved an equally startling theorem about smooth 4-manifolds.

THEOREM 3.1 [**Donaldson1**]. *If a definite form is represented by a smooth, simply connected, closed 4-manifold M, then the form is $\pm\overset{p}{\oplus}(1)$ (and by Freedman's Theorem, M is then homeomorphic to $\overset{p}{\sharp}(\pm CP^2)$).*

Thus all definite forms, with this exception, are represented by non-smoothable topological manifolds.

With a bit of further work, using ideas of Casson [**G-M**], it was shown that R^4 has more than one smooth structure (see XIV); eventually [**Taubes**] a continuum of smooth structures on R^4 was found.

In later work, Donaldson showed that if $\overset{2r}{\oplus} E_8 \overset{s}{\oplus} \begin{pmatrix} 0 & 1 \\ 1 & 0 \end{pmatrix}$ was represented by a closed, smooth simply connected 4-manifold, then $s \geq 3$ [**Donaldson2**]; in particular the Kummer surface could not split off an $S^2 \times S^2$.

Then he showed [**Donaldson3**] that the form $(1) \overset{9}{\oplus} (-1)$ is represented not only by $CP^2 \overset{9}{\sharp} (-CP^2)$ but also by the Dolgachev surface ([**Dolgachev**], [**H-K-K**]) which is obtained from the rational, elliptic surface by logarithmic transforms of multiplicity 2 and 3. Later Friedman and Morgan [**F-M1**] and Okonek and Van den Ven [**O-V**] showed that there are countably many, smoothly different, complex surfaces, obtained by logarithmic transforms of relatively prime multiplicities, with the homotopy type of $(1) \overset{9}{\oplus} (-1)$ and hence homeomorphic to $CP^2 \overset{9}{\sharp} (-CP^2)$.

A corollary to these examples is that the smooth h-cobordism "theorem" fails since these manifolds are smoothly h-cobordant ([**Wall1**] and Theorem X.1 in these notes). (The topological version is true as proved in [**Freedman1**].)

I cannot bring the reader up to date in "Donaldson theory" for new results appear frequently, especially concerning complex surfaces (see [**F-M2**]). But Fintushel and Stern's variations [**F-S1**] on Donaldson's first theorem and their work concerning whether homology 3-spheres bound smooth acyclic 4-manifolds [**F-S2**] should be mentioned; the old problem of which elements of $H_2(S^2 \times S^2; Z)$ are represented by smooth imbedded 2-spheres was settled in [**Kuga**].

IV. SPIN STRUCTURES

Let ξ be an n-plane bundle over a CW-complex X. One attractive definition of a spin structure [**Milnor2**] on ξ is the one analogous to an orientation on ξ: ξ is orientable if ξ has a trivialization over the 0-skeleton which extends over the 1-skeleton, and an orientation is a specific homotopy class of trivializations; similarly ξ can be given a spin structure if ξ has a trivialization over the 1-skeleton which extends over the 2-skeleton, and a spin structure is a homotopy class of such trivializations. However, this definition is only correct for $n \geq 3$ (the tangent bundle of an orientable 2-manifold has a spin structure, but is trivial only for the 2-torus). We can overcome the difficulty when $n \leq 2$ by stabilizing ξ. Recall that a trivialized sub k-plane bundle of a trivialized $(n+k)$-plane bundle determines a trivialization of the orthogonal n-plane bundle over the $(n-1)$-skeleton of the base space. Thus the trivializations of $\xi^n \oplus \varepsilon^k$ naturally correspond to trivializations of ξ^n over the 2-skeleton when $n \geq 3$ and ε^k is a trivialized bundle.

So we can extend the above definition: *ξ has a spin structure if $\xi \oplus \varepsilon^k$ has a trivialization over the 1-skeleton which extends over the 2-skeleton, and a spin structure is a homotopy class of such trivializations, where ε^k is a trivialized bundle.*

In practice, we do not stabilize for $n \geq 3$, add ε^1 for $n = 2$, and either add ε^2 for $n = 1$ or just add ε^1 and work "mod 2" with trivializations.

We can try a different analogy and let ξ be a principle $O(n)$-bundle. Then ξ is orientable if ξ can be reduced to an $SO(n)$-bundle, and an orientation is a given reduction. An orientation reduces the group to a connected one, $SO(n)$. To make the group simply connected, let $\text{Spin}(n) \xrightarrow{\lambda} SO(n)$ be the double cover of $SO(n)$; in particular, Spin(2) is the connected double cover of the circle $SO(2)$, and Spin(1) is $Z/2$, the only double cover of the point, $SO(1)$. Then ξ is spinnable if ξ can be covered by a Spin(n)-bundle, and a spin structure is such a covering. This can be made more precise by saying that a spin structure on ξ is a cohomology class $\sigma \in H^1(E(\xi); Z/2)$ whose restriction to each fiber is a generator of $H^1(SO(n); Z/2)$; (note that the last proviso disappears for $SO(1)$). The point is that σ determines a 2-fold covering of $E(\xi)$ and the restriction requires that the fiber $SO(n)$ be covered by its (unique if $n > 2$) double cover Spin(n). This definition is summarized by the commutative diagram:

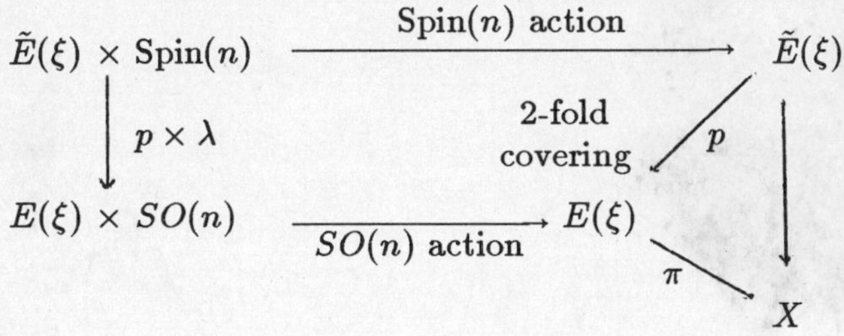

The $O(n)$-bundle ξ can be oriented iff the first Stiefel–Whitney class $\omega_1 \in H^1(X;Z/2)$ is zero, and then the orientations corresponds to $H^0(X;Z/2)$. Similarly, the $SO(n)$-bundle ξ can be given a spin structure iff the second Stiefel–Whitney class $\omega_2 \in H^2(X;Z/2)$ is zero, and then the spin structures correspond to $H^1(X;Z/2)$.

The exact sequence

$$0 \longrightarrow H^1(X;Z/2) \xrightarrow{\pi^*} H^1(E(\xi);Z/2) \xrightarrow{i^*} H^1(SO(n);Z/2) \xrightarrow{\delta} H^2(X;Z/2)$$
$$H^1(SO(n);Z/2) = Z/2$$

can be derived from a spectral sequence; $\omega_2 = \delta(1)$. Thus $\omega_2 = 0$ iff there exists a spin structure $\sigma \in H^1(E(\xi);Z/2)$ with $i^*(\sigma) = 1$, and then spin structures are classified by $\{\sigma \in H^1(E(\xi);Z/2) \mid i^*(\sigma) = 1\}$ which is isomorphic as a set to $H^1(X;Z/2)$. The spin structures are not naturally a group nor do they naturally contain a zero or preferred spin structure; however a Lie group has a natural trivialization of its tangent bundle which then gives a preferred spin structure.

That the two definitions of existence of spin structures are equivalent follows because each is equivalent to $\omega_2 = 0$ (the first by the standard interpretation of characteristic classes as obstructions to existence of cross-sections). A trivialization of ξ (or its associated principal bundle) over the 2-skeleton $X_{(2)}$ of X naturally determines a spin structure $\sigma \in H^1(E(\xi);Z/2)$. For if $\varphi : E(\xi \mid X_{(2)}) \to X_{(2)} \times SO(n)$ is a homeomorphism, and if $\sigma_0 = (0,1) \in H^1(X_{(2)};Z/2) \oplus Z/2 = H^1(X_{(2)} \times SO(n);Z/2)$ then let $\sigma = \varphi^*\sigma_0$.

PROPOSITION 1. *Suppose that ξ has a spin structure σ. Suppose that Y is a subset of X for which $H^1(X;Z/2) \to H^1(Y;Z/2)$ is an isomorphism (e.g. a basis for $H_1(X;Z/2)$). Then a spin structure on ξ is determined by the choice of a spin structure τ on $\xi \mid Y$.*

PROOF: Define a 1-cochain c on Y with values in $\pi_1(SO(n)) = Z/2$ by letting c be 0 (respectively 1) on a loop ℓ if the spin structures σ and τ on ξ and $\xi \mid Y$ agree (resp. disagree) over ℓ. Then c determines a cohomology class $\gamma \in H^1(Y;Z/2) = H^1(X;Z/2)$ and $\sigma + \gamma$ determines a spin structure on ξ which equals τ over Y. □

Yet another definition (in a sense the best) of a spin structure on ξ is a homotopy class of lifts of the classifying map f_ξ of ξ to $B\mathrm{Spin}(n)$, i.e.

$$\begin{array}{ccc} & \nearrow & B\mathrm{Spin}(n) \\ & & \downarrow \pi \\ X & \xrightarrow[f_\xi]{} & BSO(n). \end{array}$$

Then w_2 is the obstruction to lifting f_ξ, and it lies in $H^2(X;\pi_1(SO(n)/\mathrm{Spin}(n)) = H^2(X;Z/2)$; lifts of f_ξ are classified by $H^1(X;\pi_1(SO(n)/\mathrm{Spin}(n)) = H^1(X;Z/2)$, where $SO(n)/\mathrm{Spin}(n)$ is the fiber of π above.

A manifold has orientations and spin structures according to those on T_M or on its bundle of tangent frames, F_M.

EXAMPLES: An orientable 4-manifold M^4 has a spin structure if $\omega_2 = 0$; when M is simply connected, this is equivalent to having an even intersection form (recall II, §4). If M^4 is a complex surface, then ω_2 is the mod 2 reduction of the first Chern class c_1.

An orientable 3-manifold always has a spin structure because the tangent bundle is trivial ([**M-S**], exercise 12.B, and VII, Theorem 2). An orientable 2-manifold always has a spin structure because ω_2 is the mod 2 reduction of the Euler class which is even.

A circle has two spin structures; one is the trivial $Z/2$-bundle covering the trivial $SO(1)$-bundle over S^1 and the other is the connected $Z/2$-bundle (the non-trivial S^0-bundle over the circle). The first corresponds (using the only! trivialization of T_{S^1}) to $\sigma = 0$ and the second to $\sigma \neq 0$ in $H^1(E(\xi); Z/2) = H^1(S^1; Z/2) = Z/2$.

This can also be seen by considering $T_{S^1} \oplus \varepsilon^1$; this bundle is naturally trivialized by using the orientation of the circle to trivialize T_{S^1}, and this corresponds to $\sigma = 0$ and is called the Lie group spin structure. If we change this trivialization $E(T_{S^1} \oplus \varepsilon^1) \to S^1 \times R^2$ by rotating R^2 once as S^1 is traversed, then we obtain the other spin structure corresponding to $\sigma = 1$.

One analyzes the spin structures on the torus, T^2, or other surfaces in the same way.

For questions of bordism, we must relate spin structures on ∂M and spin structures on M. Given a trivialization of $T_{\partial M}|$2-skeleton, we can extend it to a trivialization of T_M on the 2-skeleton of ∂M, $(\partial M)_{(2)}$, by adding as "last vector" the inward pointing normal vector to ∂M. This coincides with the usual orientations for B^2, $\{(1,0),(0,1)\}$, and S^1 (counterclockwise), for example. Conversely, a trivialization of $T_M \mid M_{(2)}$ restricts to give a trivialization of $T_{\partial M} \oplus \varepsilon^1 \mid \partial M_{(2)}$ (where ε^1 is trivialized as the inward pointing normal vectors) which is a spin structure on ∂M according to the "stable" definition.

Alternately, consider spin structures as 2-fold covers of the total space of the frame bundle, $E(F_M)$. $F_M|\partial M$ has a subbundle $V_{\partial M}$ consisting of frames of F_M whose last vector points inward. Then a 2-fold cover of $E(F_M)$ naturally gives one for $E(V_{\partial M}) = E(F_{\partial M})$ and hence a spin structure on ∂M.

Now we say that a spin manifold M^m is a *spin* boundary of a spin manifold W^{m+1} if M is diffeomorphic to ∂W^{m+1} and the diffeomorphism carries the spin structure on M to the spin structure on W^{m+1} restricted to ∂W^{m+1}. And Ω_m^{spin} is defined to be equivalence classes of spin m-manifolds, where M_1^m and M_2^m are equivalent if M_1 and $-M_2$ together spin bound a spin W^{m+1}.

EXAMPLES: The circle is a subtle example. T_{B^2} has a unique trivialization, hence spin structure. This trivialization, when restricted to $S^1 = \partial B^2$, gives a trivialization of $T_{S^1} \oplus \varepsilon^1$ which differs from the Lie group spin structure on S^1, and hence corresponds to $1 \in H^1(S^1; Z/2)$, (see Figure 1)

This argument extends to any orientable, hence spin, 2-manifold with circle boundary. Thus the non-zero class in $H^1(S^1; Z/2)$ is the spin structure that spin bounds and the zero class does not. So $\Omega_1^{\text{spin}} = Z/2$.

The torus T^2 has four spin structures corresponding to $H^1(T^2; Z/2) = Z/2 \oplus Z/2$. As in the circle case, the Lie group framing corresponds (under the natural trivialization for $T^2 = R^2/Z^2$) to $(0,0)$ and this is the spin structure which does not bound. The others extend over a solid torus $S^1 \times B^2$, where B^2 is glued onto a circle in T^2 on which the spin cohomology class is 1.

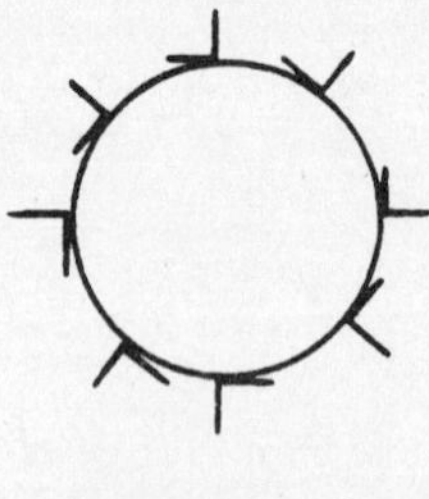

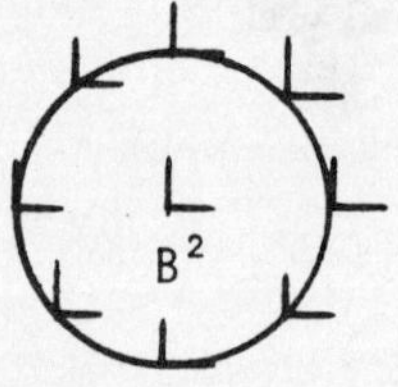

Lie group spin structure on S^1

The bounding spin structure on S^1

Figure 1

More generally, a connected sum of tori with non-Lie group framings bounds a compatible spin manifold. Also the connected sum of $2k$ tori with Lie group framings bounds $(\overset{k}{\#}\, T^2 - \text{disk}) \times I$ and the spin structure can be seen to extend to the Lie group framing. Thus we see that $\Omega_2^{\text{spin}} = Z/2$, generated by T^2_{Lie}.

We can understand this isomorphism using the Arf invariant (see the Appendix). For an oriented, spun surface F^2, consider $H_1(F^2; Z/2)$ with its usual $Z/2$ intersection form, and define a quadratic function $q : H_1(F^2; Z/2) \to Z/2$ as follows: represent $x \in H_1(F^2; Z/2)$ by an imbedded circle and let $q(x) = 0$ if the spin structure on F^2 restricted to the circle is the bounding one, and let $q(x) = 1$ if not. Thus $q = 1$ on both generators of T^2_{Lie}. Then $\text{Arf}(H_1(F; Z/2), q) \in Z/2$ gives the isomorphism $\Omega_2^{\text{spin}} \xrightarrow{\text{Arf}} Z/2$.

$\Omega_3^{\text{spin}} = 0$. This can be shown in various ways, including a handlebody argument in [**Kaplan**] and Theorem 3 of Chapter VII.

Notice that if we take the connected double cover of the circle, then both spin structures on S^1 lift to the Lie group spin structure on S^1 since the double cover induces the zero map $H^1(S^1; Z/2) \xrightarrow{0} H^1(S^1; Z/2)$. Corresponding remarks hold for 2^k-fold covers of spin structures on T^n which correspond to elements $(m_1, m_2, \ldots, m_n) \in H^1(T^n; Z/2)$ with k non-zero entries.

Our study of spin structures on M and ∂M generalize to:

PROPOSITION 2. *Let N be a codimension one submanifold of M and suppose that M, N and the normal line bundle are orientable and oriented consistently. Then there is a natural one-to-one correspondence between spin structures on $T_M \mid N$ and N, given by using the canonical spin structure on the normal line bundle.*

For other codimensions, the following can be shown:

PROPOSITION 3. *If* $\xi = \xi_1 \oplus \xi_2$, *then orientations on two of the bundles determine an orientation on the third, and similarly, spin structures on two of the bundles determine a spin structure on the third.*

Proposition 3 applies to a case that is of relevance later. Suppose F^2 is an orientable surface smoothly imbedded with a trivial normal bundle in a spin 4-manifold M^4. The trivializations of the normal bundle are classified by $H^1(F^2; Z)$, so we may obtain any spin structure on F^2 from the spin structure on N^4 and the right choice of trivialization of the normal bundle.

V. T^3_{Lie} AND $CP^2 \overset{9}{\#} (-CP^2)$

The 3-torus T^3 has 8 different spin structures corresponding to $H^1(T^3; Z/2) = (Z/2)^3$. $(0,0,0)$ is the Lie group spin structure, T^3_{Lie}. The other spin structures all spin bound either $T^2 \times B^2$ or $S^1 \times B^2 \times S^1$ or $B^2 \times T^2$ with an appropriate spin structure on T^2 crossed with the unique structure on B^2. Since $\Omega_3^{\text{spin}} = 0$, we know that T^3_{Lie} also spin bounds and we describe such a spin 4-manifold in three ways.

Let $\alpha = 3\alpha_0 + \alpha_1 + \alpha_2 + \cdots + \alpha_9 \in H_2(CP^2 \overset{9}{\#} (-CP^2); Z) = Z^{10}$ where α_i is the generator corresponding to the ith $\pm\ CP^2$. α_i is represented by a 2-sphere, CP^1, and $3\alpha_0$ is known to be represented by a torus, so α is represented by the connected sum which is also a torus T^2.

(Here is a quick proof that $3\alpha_0$ is represented by a torus. From projective geometry, $3\alpha_0$ is represented by three complex lines, three CP^1's, which can be chosen generically to meet at three points p_1, p_2, and p_3 in CP^2. A 4-ball neighborhood around p_1 intersects the two CP^1's in a pair of transverse 2-balls meeting only at p_1. If we cut out the interiors of these 2-balls and glue in an annulus, we have removed the singular point and taken the connected sum of the two CP^1's (see Figure 1). Doing the same at p_2 gives us an immersed 2-sphere with one double point at p_3. One more surgery at p_3 gives an imbedded torus.)

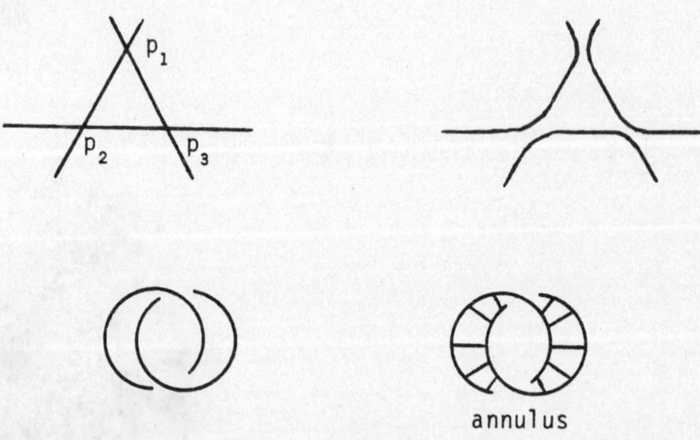

Figure 1

Our torus T^2 has a trivial normal bundle, $T^2 \times B^2$, since $\alpha \cdot \alpha = 0$. Since $\alpha \cdot x \equiv x \cdot x(2)$ for all $x \in H_2(CP^2 \overset{9}{\#} (-CP^2); Z)$, it follows that T^2 represents an integral dual to ω_2. Thus the complement of T^2 is parallelizable (see §2.6). So $N^4 = (CP^2 \overset{9}{\#} -CP^2) - (T^2 \times \text{int } B^2)$ is spin with $\partial N = T^3$. We know that the spin structure induced on T^3 is the Lie group one, because if not, we could extend the spin structure on N^4 over some $T^2 \times B^2$ added to ∂N to make a closed, smooth, spin manifold of index 8! This contradicts Rohlin's theorem (III, §1 and XI). But we wish to use this construction in work leading to our proof of Rohlin's theorem in Chapter XI, so we give another description of N^4 related to complex elliptic surfaces.

We will describe $X = CP^2 \overset{9}{\#} (-CP^2)$ as a complex elliptic surface, which, for our purposes, is a complex analytic projection $X \xrightarrow{\pi} CP^1$ which is a smooth fiber bundle with fiber T^2 (an elliptic curve) except for a finite number of singular fibers. In our case we can take these singular fibers to be immersed 2-spheres with one transverse double point, and there will be 12 of them over $p_1, p_2, \ldots, p_{12} \in CP^1$. (See the schematic diagram in Figure 2.)

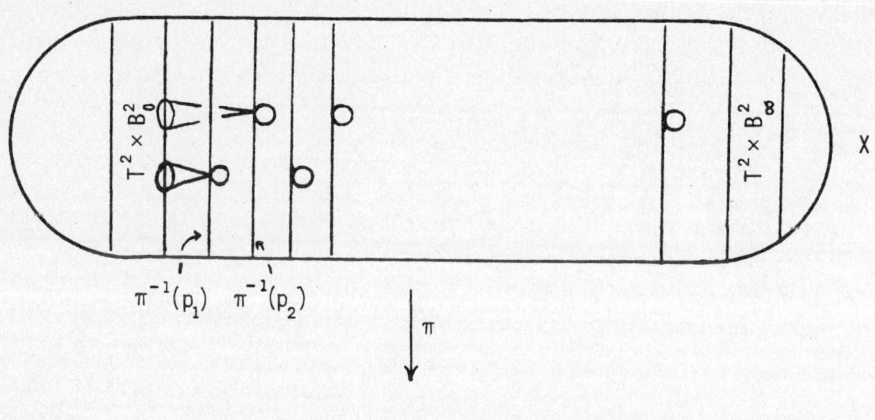

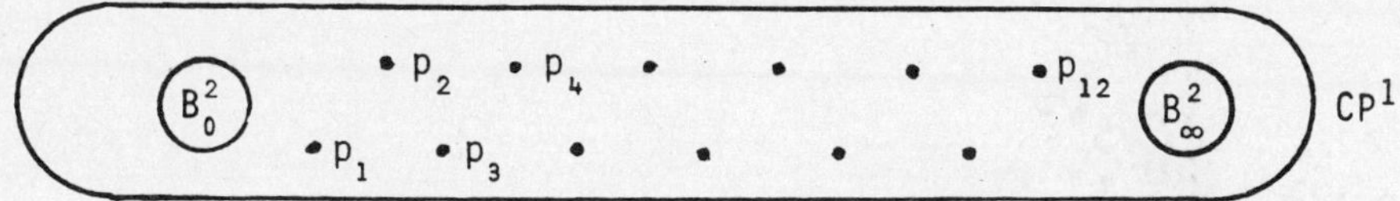

Figure 2

If we begin with the trivial bundle over $B_0^2 \subset CP^1$, $T^2 \times B_0^2$, then as the 2-ball B_0^2 expands, a fiber T^2 moves closer and closer to the singular fiber $\pi^{-1}(p_1)$ and we see a loop λ in T^2 shrink smaller and smaller until it pinches to a point (the transverse double point) in $\pi^{-1}(p_1)$. The loop λ is called a "vanishing cycle" because it vanishes homologically in X. Figure 3 shows how that might happen except that the double point is not transverse. As B_0^2 expands, reaching one p_i after another, different loops can vanish; in our case for odd p_i a meridian vanishes and for even p_i a longitude vanishes. This picture motivates the following construction.

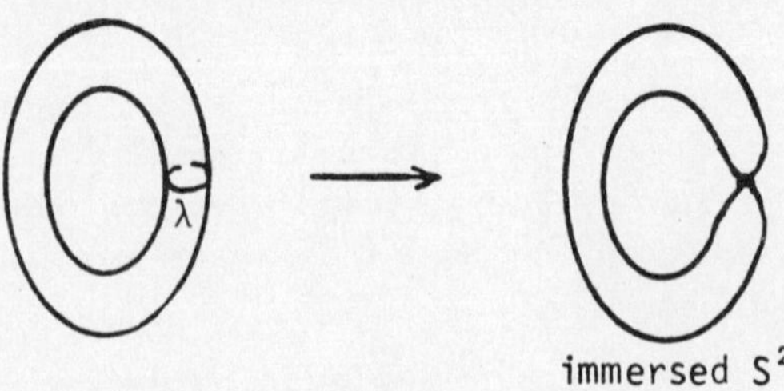

Figure 3

We begin with a handlebody description of $T^2 \times B_0^2$, which is just the usual 0, two 1's, and a 2-handle for T^2 fattened by B_0^2 (see Figure 4). To this we add a 2-handle to a meridian, say the horizontal 1-handle, with framing -1. The core of the 2-handle is the vanishing cycle for the meridian, as adding the 2-handle is the same as enlarging B_0^2 to include p_1. This manifold ought to be equal to a fattened singular fiber, which can be seen by cancelling the horizontal 1- and 2-handles as in Figure 4.

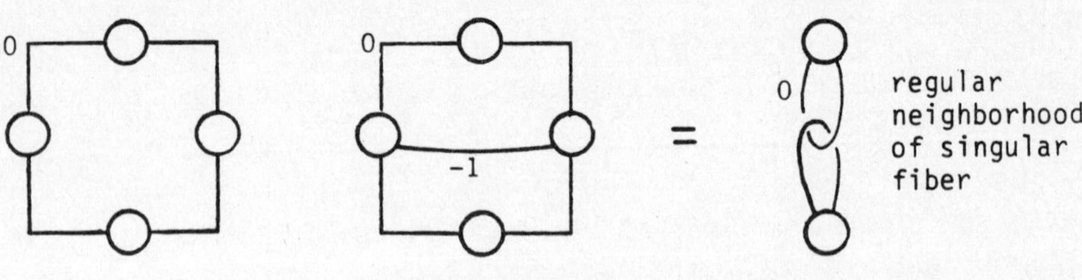

Figure 4

We continue to enlarge B^2, adding twelve 2-handles, all with framing -1, alternately to meridians and longitudes (see Figure 5). The last step is to fill in the final $T^2 \times B_0^2$, which is done by adding a 2-handle, two 3-handles, and a 4-handle. The 2-handle is added, say, in the lower left corner, again with framing -1; the 3- and 4-handles are not drawn as usual.

We can ignore the heuristic motivation, but several things need to be checked for this construction to work. By the usual handle slides, cancellations, and isotopies, it can be shown that the boundary in Figure 5, ∂M_{L_x}, is ^{0}(Hopf link)^0 which is $S^1 \times S^2 \,\#\, S^1 \times S^2$, to which it is obvious how to add two 3-handles and a 4-handle. It is somewhat harder, but a reasonable exercise (see [**H-K-K**]) to show that M_{L_x} is diffeomorphic to X^4.

But our main aim is to show that the complement of $T^2 \times B_0^2$, N^4, is a spin manifold whose spin boundary is T^3_{Lie}. We need to verify that the Lie group spin structure extends over all the 2-handles; the 3- and 4-handles are not relevant to the 2-skeleton. All the 2-handles are attached to one of $S^1 \times * \times *$ or $* \times S^1 \times *$ or $* \times * \times S^1$, so it suffices

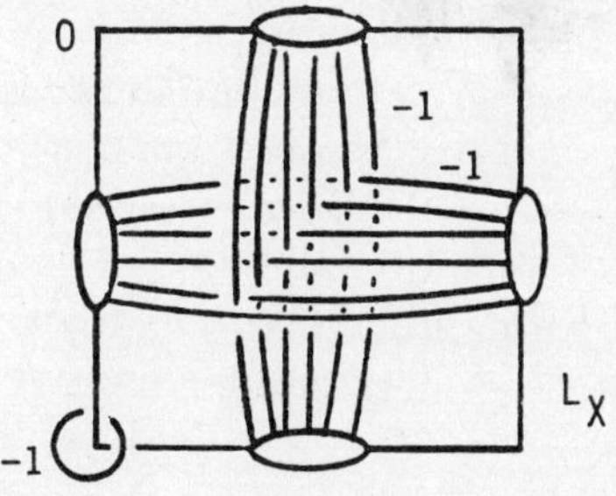

Figure 5

to examine one case, say, $* \times * \times S^1 = * \times * \times \partial B_0^2$ which is the 2-handle in the lower left-hand corner of Figure 5. The Lie group framing of T_X over this S^1 is given by the tangent vector to S^1, the inward pointing normal to ∂B_0^2 in B_0^2, and the "trivial framing" from T^2 (see Figure 6).

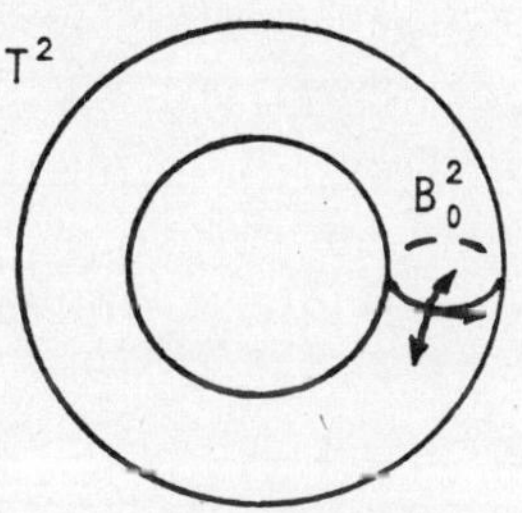

Figure 6

When we add $B^2 \times B^2$ along $\partial B^2 \times B^2$ to ∂B_0^2 we need to change the Lie group framing on ∂B_0^2 to the one which extends over $B^2 \times 0$ in the 2-handle. This can be done by adding the 2-handle with an odd framing instead of an even one; -1 works best in order to be able to add the 3- and 4-handles to cap off. The reader might reconsider Proposition 3 and the remark at the end of Chapter IV on spin structures. To restate the issue, a spin structure will extend across a 2-handle if the spin structure on the attaching circle spin bounds and thus extends across $B^2 \times O$, and the attaching framing is even, or (as in this case) if the spin structure on the attaching circle is the non-bounding one and we add the 2-handle with odd framing.

So we have constructed our spin manifold N^4 with spin boundary T^3_{Lie}, a second time.

A third description of N is this: T^3 results from surgery on the Borromean rings with 0-framings. (To see this take our picture of $T^2 \times B^2$ in Figure 4, and change the notation for 1-handles by replacing the dots with zeros (see Chapter I, §2).) The three linking circles to the Borromean rings are the generating circles in T^3; to them we add three 2-handles with framing +1, odd so that, as before, the Lie group spin structure extends over the 2-handles. We have a spin manifold N_0 consisting of $T^3 \times I$ union three 2-handles, with boundary T^3_{Lie} and the Poincaré homology 3-sphere. To prove the latter, blow down the three +1 circles to get the Borromean rings with −1 framings; blow down two of these to obtain the left-handed trefoil knot with −1 framing which describes the Poincaré homology 3-sphere with the orientation which bounds the $-E_8$ 4-manifold with index −8 (see Chapter I, §5). The spin manifold $-E_8$ has a spin boundary which must coincide with the one on N_0 because the Poincaré homology sphere has a unique spin structure ($H_1 = 0$). Thus $N^4 = N_0 \cup (-E_8)$. (Of course we leave to the reader the proof that the three descriptions of N^4 actually give the same manifold.)

VI. IMMERSING 4-MANIFOLDS IN R^6

Let M^4 be a closed, smooth, oriented, 4-manifold. If $f : M^4 \to R^6$ is a smooth immersion, then the normal bundle ν satisfies $\tau_M \oplus \nu = \varepsilon^6$. Conversely, if we can find a 2-dimensional inverse bundle ν to τ_M, then there is a bundle map $\tau_M \oplus \nu \to \varepsilon^6$ and by the immersion theorem [**Hirsch1**] there is an immersion $f : M^4 \to R^6$.

We construct ν as follows: over the 1-skeleton of M^4, τ_M is trivial so we let ν be trivial also. $\omega_2(\tau_M) \in H^2(M; \pi_1(SO(4)))$ is the obstruction to trivializing τ_M over the 2-skeleton. If $\omega_2(\tau_M)$ is the reduction of an integral class $\chi \in H^2(M; Z)$, then since $Z = \pi_1(SO(2))$ maps onto $\pi_1(SO(4)) = Z/2$ we can extend ν over the 2-skeleton of M so that its Euler class is $\chi(\nu) = \chi \in H^2(M; \pi_1(SO(2)))$. Since $\pi_2(SO(2)) = \pi_3(SO(2)) = 0$ there is no obstruction to a (unique) extension of ν over the rest of M.

Is $\tau \oplus \nu$ trivial? Since $\omega_2(\tau \oplus \nu) = \omega_2(\tau) \cup 1 + 1 \cup \omega_2(\nu) = \omega_2(\tau) + \chi_{(2)} = 0$ and $\pi_2(SO(6)) = 0$, $\tau \oplus \nu$ is trivial on M^4-point. The obstruction to trivializing $\tau \oplus \nu$ over the last point is an element of $\pi_3(SO(6))$ which is measured by p_1, as we shall see.

From the exact sequences for the fiber bundles $SO(4) \to SO(5) \to S^4$ and $SO(5) \to SO(6) \to S^5$, and from an argument below, we see that

$$\pi_3(SO(4)) \overset{\text{onto}}{\twoheadrightarrow} \pi_3(SO(5)) \cong \pi_3(SO(6)) = Z.$$

That $\pi_3(SO(4)) = Z \oplus Z$ follows from the bundle $SO(3) \to SO(4) \to S^3$ which has a cross-section given by viewing S^3 as the unit quaternions; define $\sigma : S^3 \to SO(4)$ by $\sigma(q)(q') = qq'$. Thus $SO(4) = S^3 \times SO(3)$, and σ and ρ generate $\pi_3(S^3)$ and $\pi_3(SO(3))$ respectively where $\rho : S^3 \to SO(3)$ is defined by $p(q)(q') = qq'q^{-1}$, interpreted as an orthogonal map of the quaternions which fixes the real quaternions and rotates the complementary imaginary quaternions, a copy of R^3 spanned by i, j and k.

We may think of σ and ρ as determining principal $SO(4)$ bundles or associated R^4 or S^3 bundles over S^4. (Note that $\sigma + \rho$ determines a 3-sphere bundle whose total space is homeomorphic but not diffeomorphic to S^7 [**Milnor3**], and in fact gives the generator of the group of homotopy 7-spheres $\Gamma_7 = Z/28$.)

$\sigma : S^3 \to H(1) \subset SO(4)$ determines the quaternionic Hopf bundle over S^4 just as $S^1 \to U(1)$ determines the complex Hopf bundle over S^2 and $S^0 \to O(1)$ determines the real Hopf bundle over S^1; all of these have Euler class $\chi = 1$, because each involves "one twist" in the fiber as one traverses the equator. Then $c_2(\sigma) = \chi(\sigma) = 1$ and $c_1(\sigma) \in H^2(S^4; Z) = 0$ so $p_1(\sigma) = (c_1^2 - 2c_2)(\sigma) = -2$.

On the other hand, the tangent bundle of S^4, τ_{S^4}, is $2\sigma - \rho$ (see [**Steenrod**], §23.6). Since $p_1(\tau_{S^4}) = p_1(\tau_{S^4} \oplus \varepsilon') = p_1(\varepsilon^5) = 0$ and $\chi(\tau_{S^4}) = 2$, we can calculate that $\chi(\rho) = 0$ and $p_1(\rho) = -4$. This means that bundles over S^4 are determined by their Euler class and Pontrjagin class because these give an isomorphism

$$\pi_3(SO(4)) \xrightarrow{(\chi,\ -(p_1 + 2\chi)/4)} Z \oplus Z$$

which sends σ to $(1,0)$ and ρ to $(0,1)$.

In the sequence $\pi_4(S^4) \to \pi_3(SO(4)) \twoheadrightarrow \pi_3(SO(5))$, $1 \in \pi_4(S^4)$ hits $2\sigma - \rho = T_{S^4} \in \pi_3(SO(4))$, so it follows that the stabilization of ρ, $\rho \oplus \varepsilon^1$, is twice the stabilization of σ, $\sigma \oplus \varepsilon^1$, in $\pi_3(SO(5))$. Thus $\sigma \oplus \varepsilon^1$ generates $\pi_3(SO(5))$; then $\sigma \oplus \varepsilon^2$ generates $\pi_3(SO(6))$ and an isomorphism with Z is given by $-p_{1/2}$.

Returning to $\tau_M \oplus \nu$ we calculate $p_1(\tau \oplus \nu) = p_1(\tau) + p_1(\nu) = p_1(\tau) + \chi^2(\nu)$ (here $p_1(\nu) = -c_2(\nu \otimes \mathbb{C}) = -\chi(\nu \otimes \mathbb{C}) = -\chi(-\nu \oplus \nu) = \chi^2(\nu)$ [**M-S**, pg. 179]), so $\tau \oplus \nu$ is trivial if $\chi^2(\nu) = -p_1(\tau_M)$. We have shown

LEMMA 1. *A smooth, closed, orientable 4-manifold M^4 immerses in R^6 iff there exists a characteristic class $\chi \in H^2(M^4; Z)$ such that $\chi_{(2)} = -\omega_2(\tau_M)$ and $\chi^2 = -p_1(\tau_M)$. Then the normal bundle ν of the immersion is determined by $\chi(\nu) = \chi$.*

COROLLARY 2. *If M^4 is spin and $p_1(\tau_M) = 0$, then M immerses in R^6 with trivial normal bundle (i.e. $\chi = 0$).*

Note that M actually immerses in R^5 with trivial normal bundle since $\tau_M \oplus \varepsilon^1$ is also trivial because $\omega_2(\tau_M) = p_1(\tau_M) = 0$.

COROLLARY 3. *If $\pi_1(M^4) = 0$,* index$(M^4) = 0$ *and $p_1(\tau_M) = 0$, then M immerses in R^6.*

PROOF: We need an integral characteristic class with square zero. The intersection form is indefinite and we may assume, because Corollary 2 covers the even case, that it is odd. Since the index is zero, the form is $\overset{k}{\oplus} \langle 1 \rangle \overset{k}{\oplus} \langle -1 \rangle$, so our characteristic class can be chosen to be the sum of the generators of the form. □

Note that we have not yet proven the index theorem (Chapter IX), so we assume both index $M = 0$ and $p_1(\tau_M) = 0$.

Next we show that the number of triple points of an immersion $f : M^4 \to R^6$, counted algebraically, is $-p_1(M)/3$, (see [**Herbert**], pg. xiii, Coro. 6). Let Δ^2 be the set of double points of f, i.e., let $\Delta^2 = \{x \in M^4 \mid \exists y \in M^4 \text{ with } f(x) = f(y)\}$; Δ^2 is an immersed, oriented 2-manifold in M which double covers its image $f(\Delta)$. The orientation of Δ is determined by the orientation of M, which orients the normal bundle of M, which orients the normal bundle to Δ in M, which orients Δ.

LEMMA 4. *The homology class $[\Delta] \in H_2(M; Z)$ is the Poincaré dual to $-\chi(\nu)$ and an integral dual to $\omega_2(\tau_M)$.*

PROOF: This is just intersection theory; perturb $f(M)$ to M' transverse to $f(M)$. Then $[f(M)] = 0 \in H_4(R^6; Z)$, implies that $[f(M) \cap M'] = 0 \in H_2(M; Z)$ (since a 5-chain with boundary $f(M)$ intersected with M' gives a 3-chain with boundary $f(M) \cap M'$). Then it follows that $0 = [f(M) \cap M'] = [\Delta] + P.D.(\chi(\nu))$ where the second equality is obtained by counting the contributions from $f(\Delta)$, namely $[\Delta]$, and from the twisting in the normal bundle, namely the Poincaré dual to $\chi(\nu)$ (see Figure 1 for a low dimensional example). □

Let $\sharp M$ be the algebraic number of triple points of $f(M)$ in R^6. (Each occurs with a plus or minus sign according to whether the three oriented normal planes combine to

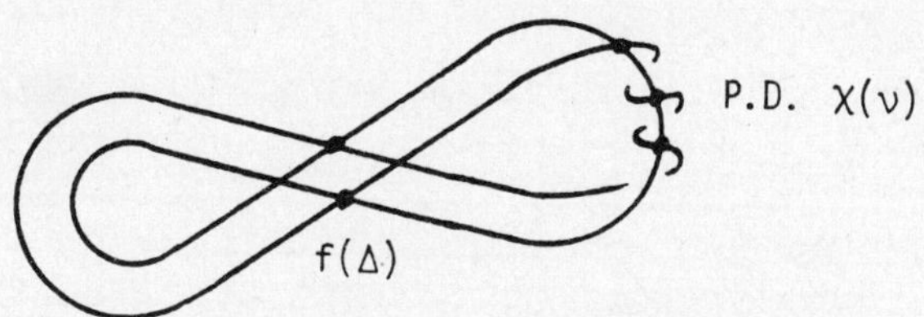

Figure 1

give the orientation of R^6 or not.) Let $\sharp\Delta$ be the algebraic number of double points of Δ in M. Clearly $\sharp\Delta = 3\sharp M$.

In the case that M^4 is spin (and $p_1(M^4) = 0$), then M immerses in R^5, and the 2-spine of M imbeds in R^5. Assuming that M has no 3-handles (M is bordant to such a manifold), we can push the interior of the 4-handle off R^5 into R^6 so that Δ lies in the 4-handle. Therefore $[\Delta] = 0$ and the normal bundle of Δ in M is trivial, so $\sharp\Delta = 0 = \sharp M$. A bit more sophisticated argument proves

LEMMA 5. *$-p_1(M) = [\Delta]\cdot[\Delta] = \chi(\nu_\Delta)[\Delta] + 2\sharp\Delta = 3\sharp M$ where ν_Δ is the normal bundle of Δ in M^4.*

PROOF: The first equality follows from applying

$$-p_1(\tau_M) = \chi(\nu)^2 = -P.D.[\Delta] \cup -P.D.[\Delta]$$

to the fundamental class of M. The second equality is just intersection theory again, as in the proof of Lemma 4. (Note that we have $2\sharp\Delta$ because $\sharp\Delta$ counts the points in the image of the double point set of Δ.) The third equality follows from the fact that under $H^2(M;Z) \to H^2(\Delta, Z)$, we have $\chi(\nu_M) \to \chi(\nu_\Delta)$, so

$$\chi(\nu_\Delta)[\Delta] = \chi(\nu)[\Delta] = (-P.D.[\Delta])[\Delta] = -[\Delta]\cdot[\Delta].$$

Thus $[\Delta]\cdot[\Delta]$ equals both $-\chi(\nu_\Delta)[\Delta]$ and $\chi(\nu_\Delta)[\Delta] + 2\sharp\Delta$, so $\chi(\nu_\Delta)[\Delta] = -\sharp\Delta$ and that implies that $\chi(\nu_\Delta)[\Delta] + 2\sharp\Delta = \sharp\Delta = 3\sharp M$. □

VII. 3-MANIFOLDS; A DIGRESSION

The methods used in the next section, to show that a spin 4-manifold with index zero spin bounds a spin 5-manifold, also show with much less trouble that $\Omega_3^{\text{spin}} = \Omega_3^{SO} = 0$. We need a few of these facts later so we give a brief treatment of the triviality of the tangent bundle of an orientable 3-manifold, the nullity of bordism, and an easy application to imbedding orientable 3-manifolds in R^5.

THEOREM 1. *Every orientable 3-manifold M^3 is spin and hence parallelizable.*

PROOF: First, let M^3 be compact. We can assume that M^3 is closed (otherwise double M). Assume $\omega_2(M) \neq 0$ and let C be a circle in M which is Poincaré dual to $\omega_2(M) \in H^2(M^3; Z/2)$. Then $M^3 - C$ has a spin structure σ which does not extend over C. There is a dual surface F^2, perhaps non-orientable, in M^3 which intersects C transversally at one point p. The total space of the normal B^1-bundle to F^2, $F \tilde{\times} B^1$, is equal to the total space of the normal B^1-bundle to an immersion of F in R^3 (which always exists). Therefore $F \tilde{\times} B^1$ has a spin structure. The spin structures on $F \tilde{\times} B^1$ are classified by $H^1(F; Z/2)$ which equals $H^1(F-p; Z/2)$ which classifies spin structures on $(F-p) \tilde{\times} B^1$. Then σ gives a spin structure to $(F-p) \tilde{\times} B^1$ which must agree with one on $F \tilde{\times} B^1$. Thus σ extends across C, contradicting $\omega_2(M) \neq 0$. Hence M is spin. Since $\pi_2(SO(3)) = 0$, M is parallelizable.

Suppose M^3 is non-compact. If T_M is non-trivial, then it is non-trivial on some compact piece of M^3, contradicting the above case. □

THEOREM 2. *Every orientable 3-manifold M^3 bounds an orientable 4-manifold W^4. If M^3 is connected, W^4 can be assumed to have only one 0-handle and some 2-handles.*

PROOF: Lickorish's proof [**Lickorish**] via Dehn twists and Heegaard splittings is a useful one; more recently Rourke's argument [**Rourke**] is short and elementary. The proof we sketch here is fairly easy if one has already understood the similar (but harder) proof in the next section that an orientable 4-manifold that immerses in R^6 bounds an orientable 5-manifold; or this proof can be considered as a warm up for the proof in the next section.

Since M^3 is parallelizable, it immerses in R^5 with a trival normal bundle ν. The double point set is a collection of circles in M^3 which double cover their images by either connected or disconnected double covers. These circles have trivial normal bundles in M^3, which are also the restriction of ν to the circles. Then the normal B^4 bundle ν_C to each circle C in R^5 is locally split as a Whitney sum of the two trivial B^2 bundles.

Consider the case of a disconnected double cover, i.e. two circles C_1 and C_2 whose image is C in R^5; then, choosing trivializations $C_i \times B^2$ for ν_i, the normal bundle of C_i in M^3, $i = 1,2$, we get a trivialization $C \times B^4$ for ν_C where $C_1 \times B^2$ contributes the first two factors of B^4 and $C_2 \times B^2$ the other two factors. For any point $p \in C$, we can replace the two copies of $p \times B^2$ in $p \times B^4$ by an annulus $A = S^1 \times [1,2]$ where $S^1 \times 1$ is attached to $p \times \partial B^2$ (in $C_1 \times B^2$) and $S^1 \times 2$ to $p \times \partial B^2$ (in $C_2 \times B^2$) so that the orientations match up. This annulus A can be chosen in a canonical way (see the details in the proof of Theorem 1 in Chapter 8) so this construction can be carried out in $C \times B^4$; as a result, $C_1 \times B^2$ union $C_2 \times B^2$ is removed from M and replaced by $C \times A$, giving a new 3-manifold M'. M' is obviously immersed in R^5 with one less double curve. Furthermore, M' is oriented bordant to M via a bordism consisting of $M \times [0,1]$ with a "handle" $[1,2] \times (S^1 \times B^2)$ attached by $\partial[1,2] \times (S^1 \times B^2)$ to $C_1 \times B^2$ union $C_2 \times B^2$.

The case of a connected double cover $C_1 \to C$ can be handled in a similar fashion (it is done in detail in Theorem VIII). Hence all the double curves can be eliminated and M is bordant to a 3-manifold which is imbedded in R^5; it has a "Seifert" 4-manifold (Theorem VIII.3) and the union of these two 4-manifolds is W^4.

If M is connected, W can be assumed connected and all but one 0-handle can be cancelled by 1-handles; then all remaining 1 and 3-handles can be changed to 2-handles by the construction preceding Lemma 2.1 in Chapter I.

THEOREM 3. *If M^3 is spin then M^3 spin bounds a spin 4-manifold with only 0-handles (one for each component) and 2-handles.*

PROOF: Assuming without loss of generality that M^3 is connected, we know from the previous theorem that M^3 bounds a W^4 with one 0-handle and some 2-handles. The obstruction to extending the spin structure on M over W is a closed surface F^2 in W which is dual to $\omega_2 \in H^2(W, M; Z/2)$. We can see F^2 by considering W^4 as a handlebody built on M^3; 2-handles are added and then a 4-handle. The spin structure on M^3 will not extend over some of the 2-handles, which means it will not extend past the cocores of these 2-handles. These cocores are just the cores of the original 2-handles of W; if we let the attaching circles of these cores bound an oriented Seifert surface in the 0-handle, then the Seifert surface union the cores gives an oriented F^2. Thus we see that ω_2 could have been assumed integral.

M^3 is spin bordant (via $W' = W^4 - B^2 \tilde{\times} F^2$) to the orientable circle bundle C over F^2; C has a spin structure and choosing an orientation for a normal circle chooses one for F^2, and vice versa. Suppose the Euler characteristic of C is k. By connected summing W^4 with k copies of $-CP^2$ (meaning $-k$ copies of CP^2 if $k < 0$), and connected summing F^2 with the CP^1's, we can arrange that the Euler characteristic of F^2 is zero, so that C is the trivial bundle.

Choose a section of C so that we have a homeomorphism $C \cong F^2 \times S^1$ and the spin structure on C descends to a spin structure on F^2, and a spin structure on S^1 which is the non-bounding one. Either F^2 spin bounds (the easy case) or F^2 can be written as a connected sum of an F_1 which spin bounds and a T^2 which doesn't (the harder case which we assume). Let F_1 spin bound N^3 and let T^3_{Lie} spin bound Y^4. Let W_1^4 be the "connected sum" of $N^3 \times S^1$ and Y^4 by gluing $N^3 \times S^1$ to Y^4 along the $B^2 \times S^1$ in $F_1 \times S^1$ and the $B^2 \times S^1$ in T^3_{Lie} where the B^2's in F_1 and T^2 were the ones used to form the connected sum $F = F_1 \,\sharp\, T^2$. Note that the spin structures on the two copies of $B^2 \times S^1$ coincide. ∂W_1 is spin homeomorphic to C so M spin bounds $W' \cup_C W_1$.

Since $W' \cup_C W_1$ is orientable, we can change its 1-handles to 2-handles in the usual way (see Lemma I.2.1), and by inverting the 4-manifold and using the same reasoning we can change 3-handles to 2-handles. □

THEOREM 4 [**Hirsch2**]. *Every orientable 3-manifold M^3 imbeds in $\overset{k}{\#} S^2 \times S^2$, hence in R^5.*

PROOF: By the previous theorem, M^3 bounds a spin manifold W^4 which has only one 0-handle and 2-handles. W^4 can be described by a framed link L in S^3 with even framings (see II, §4). The double of W, DW, can be described by adding a trivial linking circle with framing zero to each component of L (see Lemma I.2.5). It's not hard to see, using Lemmas I.4.4 and I.4.5 that DW has a framed link consisting of copies of (Hopf link with framings 0, 0), so $DW = \overset{k}{\#} S^2 \times S^2$. Thus DW imbeds in R^5, so M^3 also imbeds. □

REMARK: This shows that the given trivialization of τ_M over the 2-skeleton of M^3 extends to a trivialization of τ_W over the 2-skeleton of W^4 where M^3 spin bounds W^4. Since $\pi_2(SO(3)) = 0$, this fact is true for the 3-skeletons of M and W, but a trivialization of τ_M will not necessarily extend to τ_W over all of W^4.

For n large, $\pi_{n+3}(S^n) = Z/24$, and non-zero elements are represented by parallelized 3-manifolds which don't bound parallelized 4-manifolds (with the trivializations agreeing on the boundary).

VIII. BOUNDING 5-MANIFOLDS

THEOREM 1. *Let M^4 be closed, smooth, connected, and orientable.*

(A) *If* $p_1(M) = \text{index}(M) = 0$, *then there exists a smooth 5-manifold W^5 with $\partial W^5 = M^4$.*

(B) *If M is spin and $p_1(M) = 0$, then there exists a smooth, spin W^5 and $\partial W = M$ as spin manifolds.*

It follows that the 4-dimensional spin bordism group Ω_4^{spin} is Z, but the generator will not be clear until later.

This section is devoted to the proof of Theorem 1. From the hypotheses we see that M is immersed in R^6, $f : M \to R^6$, with algebraically zero triple points, according to Lemmas VI.1 and VI.5.

The first step is to cancel the triple points by changing M by a bordism. The second step is to remove the double points by another bordism so that the new 4-manifold M is imbedded in R^6. Then M bounds a 5-manifold W in R^6, and the non-spin case is finished. For part (B), it is necessary to modify W so that W is spin and M is a spin boundary. The details follow.

In the proof of Theorem 1 we will need the following theorems which are useful more generally.

THEOREM 2. *If an oriented m-manifold M^m is smoothly imbedded in an oriented $m+2$-dimensional manifold Q^{m+2} and M represents $0 \in H_m(Q^{m+2}; Z)$, then it has a trivial normal bundle ν.*

PROOF: Since the normal bundle $\begin{array}{c} E(\nu) \\ \downarrow \pi \\ M \end{array}$ is oriented, it is enough to find a non-zero cross-section; for this it suffices to show that the Euler class, $\psi(\nu)$, is zero. A classical construction in complex algebraic geometry concerning divisors is useful here; pull the normal bundle of M back over itself as in the diagram

$$\begin{array}{ccc} E(\pi^*\nu) & \longrightarrow & E(\nu) \\ s \uparrow\downarrow & & \downarrow \pi \\ E(\nu) & \xrightarrow{\pi} & M \end{array}$$

and observe that $\pi^*\nu$ has a section s defined by $s(e) = (e, e)$ for $e \in E(\nu)$. This section is non-zero on $E(\nu) - M$, so the oriented bundle $E(\pi^*\nu)$ is trivial off M. Then we can extend $\pi^*\nu$ to all of Q^{m+2} by the trivial bundle; call the extension ξ. Of course, the Euler class of ξ restricts to $\chi(\nu)$, but $\chi(\xi)$ is Poincaré dual to $[M] = 0$, so $\chi(\nu) = 0$.

□

THEOREM 3. *If an oriented, connected, m-manifold M^m is smoothly imbedded in an oriented $(m+2)$-manifold Q^{m+2} with $[M] = 0 \in H_m(Q^{m+1}; Z)$ (e.g., any $M^m \to S^{m+2}$), then M^m bounds an oriented "Seifert" manifold W^{m+1} in Q^{m+2}. (If Q^{m+2} is spin, then W is spin so M is spin, but this spin structure may not agree with a given spin structure on M.)*

PROOF: Let N be an open tubular neighborhood of M in Q. By the previous theorem, M has a trivial normal bundle, so N is diffeomorphic to $M \times R^2$. However, the trivializations of the normal bundle are classified by $H^1(M; Z)$.

Let D^2 be a normal disk to M in Q. If ∂D or even a multiple $n\partial D$ bounds homologically in $Q - N$, then $\{nD$ union this homology$\}$ intersects $[M] = 0$ non-trivially, which is impossible. Thus ∂D generates $Z \subset H_1(Q - N; Z)$; let $\alpha \in H^1(Q - N; Z)$ be its dual. Then the inclusions of ∂D^2 into $\partial(Q - N) = M \times S^1$ into $Q - N$ induce

$$\begin{array}{ccccc} H^1(Q-N;Z) & \xrightarrow{i^*} & H^1(\partial(Q-N);Z) & \xrightarrow{j^*} & H^1(S^1;Z) \\ \downarrow\cong & & \downarrow\cong & & \downarrow\cong \\ [Q-N;S^1] & \xrightarrow{\text{restriction}} & [\partial(Q-N);S^1] & \xrightarrow{\text{restriction}} & [S^1,S^1] \end{array}$$

Then $\beta = i^*\alpha$ corresponds to a homotopy class in $[\partial(Q - N); S^1]$ which can be seen to be represented by a map f_β which factors

$$\begin{array}{ccc} \partial(Q-N) & \xrightarrow{f_\beta} & S^1 \\ h\searrow & & \nearrow p_2 \\ & M \times S^1 & \end{array}$$

where h is a diffeomorphism coming from a trivialization of N. It follows that α is represented by $f_\alpha : Q - N \to S^1$ which restricts to $p_2 h$ on $\partial(Q - N)$. Make f_α transverse, rel ∂, to a point $p \in S^1$, and then $W^{m+1} = f_\alpha^{-1}(p)$ is a smooth, oriented manifold with $\partial W^{m+1} = M \times p = M$. □

PROOF OF THEOREM 1 (A): First we make M 1-connected by adding 2-handles to $M \times I$ along circles $\gamma \times 1$ which are constructed from the 1-handles of M. This gives a bordism W_1 from M to M_1. If M is spin, its spin structure extends over W_1 if we add the 2-handles with the right framing. Notice that M_1 has no 1-handles and, by the same construction upside down, no 3-handles From the hypotheses we see that M_1 can be immersed in R^6, $f_1 : M_1 \to R^6$, with algebraically zero triple points since $p_1(M_1) = 0$, according to Lemmas VI.6.1 and VI.6.5.

We will need the following two lemmas:

LEMMA 5. *M_1 is bordant to M_2 which is immersed, $f_1 : M_2 \to R^6$, with no triple points. The bordism W_2 is formed by adding 1-handles to $M_1 \times [0,1]$.*

LEMMA 6. *M_2 is bordant to M_3 (via W_3) which is imbedded, $f_3 : M_3 \to R^6$.*

Before proving the lemmas, we observe that the proof of Theorem 1 (A), the non-spin case, is finished since M bounds $W_1 \cup W_2 \cup W_3 \cup W_4$ where W_4 is the Seifert 5-manifold for M_3 guaranteed by Theorem 3.

PROOF OF LEMMA 5: M_1 is immersed in R^6 with algebraically zero triple points and double point set Δ_1. To "cancel" two triple points of opposite sign, q_0 and q_1, we choose an arc γ in $f_1(M_1)$ connecting q_0 and q_1. Assume for now that γ lies in $f_1(\Delta_1)$. We can assume γ passes through no other triple points since they have codimension 2 in $f_1(\Delta_1)$.

γ has a B^5 normal bundle in R^6. We can choose a basis b_0, b_1, b_2, b_3, b_4 for B^5 at q_0 so that b_0 is tangent to $f_1(M_1)$, b_0, b_1, b_2 are tangent to the third sheet Q_3 which is transverse to γ. See Figure 1. This basis can be extended across the normal bundle of γ, splitting it into subbundles tangent to Q_1 and Q_2 (since $\gamma_1 \cap Q_2$). Then at q_1, b_1, b_2, b_3, b_4 is a basis for the tangent space to Q'_3. Since the triple points have opposite sign, exactly one of the bases for T_{Q_3} at q_0 or $T_{Q'_3}$ at q_1 must not agree with the orientation on M_1.

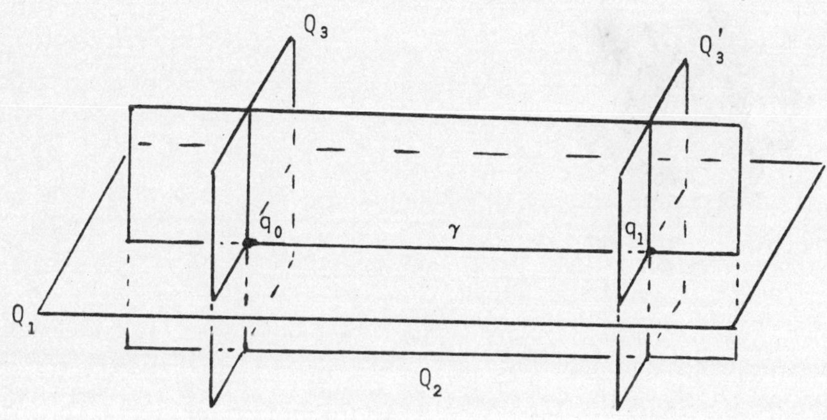

Figure 1

Then we construct the first piece of an oriented 5-dimensional bordism W_2 by adding a 1-handle $B^1 \times B^4$ to $M_1 \times I$. The 1-handle is essentially $\gamma \times B^4$ where B^4 is spanned by b_1, b_2, b_3, b_4 and it is attached to the 4-ball neighborhoods of q_0 in Q_3 and q_1 in Q'_3. The new boundary $M_1 - (q_0 \times B^4) - (q_1 \times B^4) \cup \gamma \times \partial B^4$ is immersed with two less triple points. We iterate this procedure to remove all triple points.

This process depends on $f_1(\Delta_1)$ being connected so that we can choose γ to lie in $f_1(\Delta_1)$. If not, then first we must connect components of $f_1(\Delta_1)$. Let λ be an arc in $f_1(M_1)$ joining two components of $f_1(M_1)$, as in Figure 2, with $\partial\lambda = p_0 \cup p_1$. λ should lie in one sheet Q_1 of $f_1(M_1)$ and at $\partial\lambda$ should meet sheets Q_2 and Q'_2 transversely. The normal B^5-bundle of λ has a basis b_0, b_1, b_2, b_3, b_4 at p_0 with b_0, b_1, b_2 tangent to Q_1, b_1, b_2 tangent to $Q_1 \cap Q_2$, and b_1, b_2, b_3, b_4 tangent to Q_2. Extend this splitting across λ so that p_1, b_1, b_2 is tangent to $Q_1 \cap Q'_2$ and b_1, b_2, b_3, b_4 is tangent to Q'_2 at p_1. b_1, b_2, b_3, b_4 should agree with the orientation of M at p_0 and disagree at p_1, or vice versa. If this is not true, then near q_1, γ should have been chosen to "go around" Q'_2

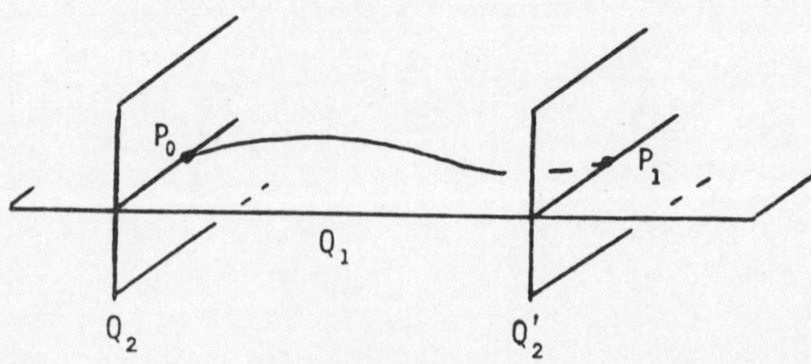

Figure 2

and approach it from the other side; this is always possible in codimension 2, although not in the dimension of Figure 2.

Then form a bordism, as above, by adding a 1-handle $\lambda \times B^4$ to $M_1 \times I$; this connects two components of $f_1(\Delta_1)$.

Thus W_2 is formed by adding a succession of 1-handles to $M_1 \times I$, first to connect the components of $f_1(\Delta_1)$ and second to remove all triple points. □

PROOF OF LEMMA 6: From Lemma 5 we get an immersion $f_2 : M_2 \to R^6$ without triple points, so it follows that the double point set Δ_2 is imbedded in M^4. From Lemma VI.5 we get $0 = 3\sharp M_2 = \sharp\Delta_2 = \chi(\nu_{\Delta_2})[\Delta_2]$ so the normal bundle of Δ_2 in M_2 is trivial if Δ_2 is connected.

The imersion $f_2 : \Delta_2 \to f(\Delta_2)$ is either a connected or disconnected double cover since $f(\Delta_2)$ is connected; we consider the connected case first, and assume that a trivialization $\Delta_2 \times B^2$ has been chosen for the normal B^2-bundle to Δ_2 in M_2. The normal B^4-bundle to $f_2(\Delta_2)$ is trivial (because its Whitney sum with the (even Euler class) tangent bundle to $f_2(\Delta_2)$ is trivial), and locally it splits as a $B^2 \times B^2$ bundle where the B^2's are given by the two sheets of $\Delta_2 \times B^2 \subset M_2$ which intersect at $f_1(\Delta_2)$. This does not split the B^4-bundle globally since the two factors switch after traversing a circle whose cover is connected.

Pick a point $q \in f_2(\Delta_2)$ and consider its normal fiber $B^2 \times B^2$. On $S^3 = \partial(B^2 \times B^2)$ we see two oriented linking circles; we want to replace $B^2 \times 0$ and $0 \times B^2$ in M_2 by an annulus $A = S^1 \times I$, in analogy with the construction in dimension 2 shown in Figure 3.

Think of B^4 as the unit ball in $\mathbb{C}^2$, and the linking circles as $(\mathbb{C}\times 0)\cap S^3$ and $(0\times\mathbb{C})\cap S^3$ and parametrize the circles by $(e^{i\theta}, 0)$ and $(0, e^{i\phi})$, $\theta, \phi \in [0, 2\pi]$. Join $(e^{i\theta}, 0)$ to $(0, e^{-i\theta})$ by the shorter component of the great circle determined by the intersection of S^3 with the real 2-plane determined by $(0,0)$, $(e^{i\theta}, 0)$ and $(0, e^{-i\theta})$. This defines, in a canonical way, an annulus A embedded in B^4 which is independent of the order of the factors $B^2 \times 0$ and $0 \times B^2$. Thus there is an A-bundle over $f_2(\Delta_2)$ imbedded in the B^4-bundle over $f_2(\Delta_2)$. We delete the intersection of M^4 with the B^4-bundle over $f_2(\Delta_2)$ from M^4 and glue in the A-bundle over $f_2(\Delta_2)$. The resulting 4-manifold is M_3 and it is smoothly (after corners are rounded) imbedded in R^6 by $f_3 : M_3 \to R^6$. M_3 is clearly orientable since each annulus A was attached to $S^1 \times 0$ and $0 \times S^1$ with an orientation preserving map on one end and reversing on the other.

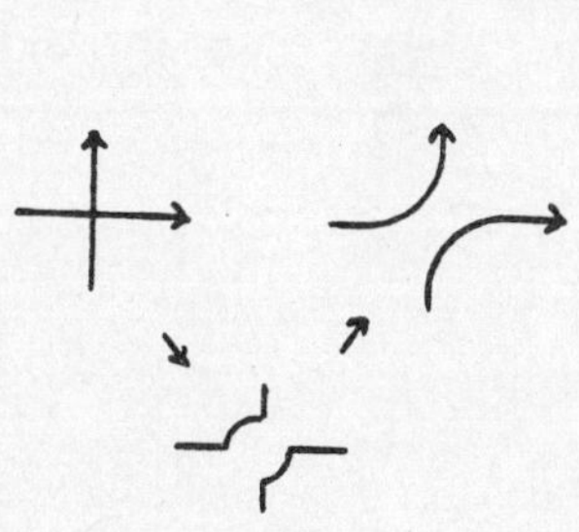

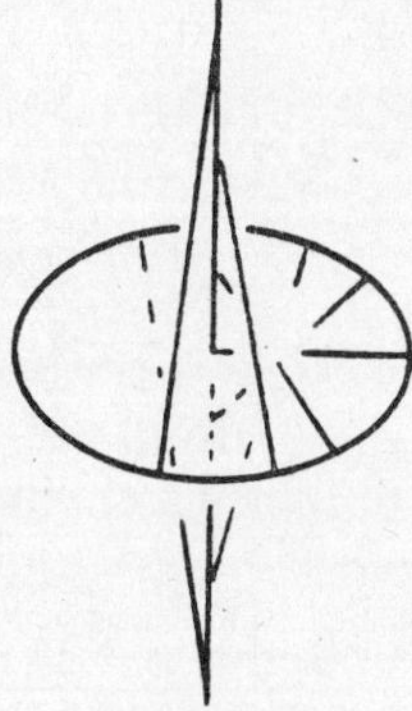

Figure 3

Next we must construct the bordism W_3 between M_2 and M_3. Assume that $\Delta_2 = T^2$ and $f_2 = (id \times z^2) : S^1 \times S^1 \to S^1 \times S^1 = f_1(\Delta_2)$ where z^2 denotes the connected double cover.

At q the bordism is just $B^1 \times B^2$ where $\partial B^1 \times B^2$ is attached to $B^2 \times 0$ and $0 \times B^2$ and the annulus A is just $B^1 \times \partial B^2$. But the whole bordism has a twist in it (because we are in the case of a connected double cover); we add to $M_2 \times I$ a "handle" H which is

$$S^1 \times (S^1 \tilde{\times} B^1) \tilde{\times} B^1 \times B^1.$$

One annulus A is $* \times * \times B^1 \times \partial(B^1 \times B^1)$. The "handle" H is attached to $M_2 \times 1$ along $\Delta_2 \times B^2 = T^2 \times B^2$ by $S^1 \times (S^1 \tilde{\times} S^0) \tilde{\times} B^1 \times B^1 = T^2 \times B^2$. The "handle" H consists of the handles of T^2 crossed with B^1 and thickened by B^2. Thus there is one 1-handle, two ($2g$ in general) 2-handles, and one 3-handle in H.

The case when genus $\Delta_2 > 1$ is similar to the case above; the "handle" H is twisted only over circles in Δ_2 which double cover their image, so we have done the prototype of the hard case and leave the rest to the reader.

Finally, consider the case of the disconnected double cover $\Delta_2 = \Delta_2' \cup \Delta_2'' \xrightarrow{f_2} f_2(\Delta_2)$ where it is possible that the two components have non-trivial normal bundles in M_2 with Euler classes of opposite sign. Choose $p \in f_2(\Delta_2)$ and a neighborhood U_α of p, let $U_\beta = f_2(\Delta_2) - p$, let U_α', U_β', U_α'', and U_β'' be "lifts" in Δ_2' and Δ_2'', and choose trivializations $U_\alpha' \times B^2$, $U_\beta' \times B^2$, $U_\alpha'' \times B^2$ and $U_\beta'' \times B^2$ so that the transition functions $g_{\alpha\beta}' : U_\alpha' \cap U_\beta' \to SO(2)$ and $g_{\alpha\beta}'' : U_\alpha'' \cap U_\beta'' \to SO(2)$ satisfy $g_{\alpha\beta}'(q') = -g_{\alpha\beta}''(q'')$ if $f_2(q') = f_2(q'')$ where the "minus" means rotation in the opposite direction.

Then the construction above for removing $B^2 \times 0$ and $0 \times B^2$ and gluing in an annulus A using the parameterizations $(e^{i\theta}, 0)$ and $(0, e^{-i\theta})$ is invariant under the action by $g_{\alpha\beta}'$ and $g_{\alpha\beta}''$ so the construction is the same over $U_\alpha \cap U_\beta$ no matter which trivializations are used, $U_\alpha' \times B^2$ and $U_\alpha'' \times B^2$ or $U_\beta' \times B^2$ and $U_\beta'' \times B^2$. This finishes the proof of

Theorem 1 (A). □

Remark: It is necessary to consider the case of an immersion $M^4 \to R^6$ in which the double point set double covers its image by a connected double cover, e.g. the double cover $S^1 \xrightarrow{2} S^1$ crossed with S^1.

Here is such an example, due to John Hughes (see [**Hughes**]), of $S^4 \to R^6$ with the torus double covering its image. First consider the immersion of S^2 into $R^4_+ = \{x \in R^4 \mid x_4 \geq 0\}$ with one double point, as drawn in Figure 4: we see a collection of slices of S^2 obtained by fixing x_3 in R^4_+, with the south pole of S^2 at $(0,0,-1,1)$ and the north pole at $(0,0,1,1)$ and the double point at $(0,0,0,1)$.

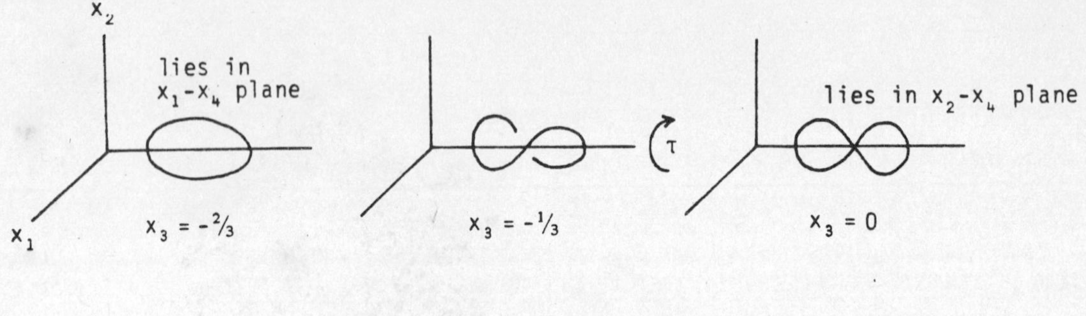

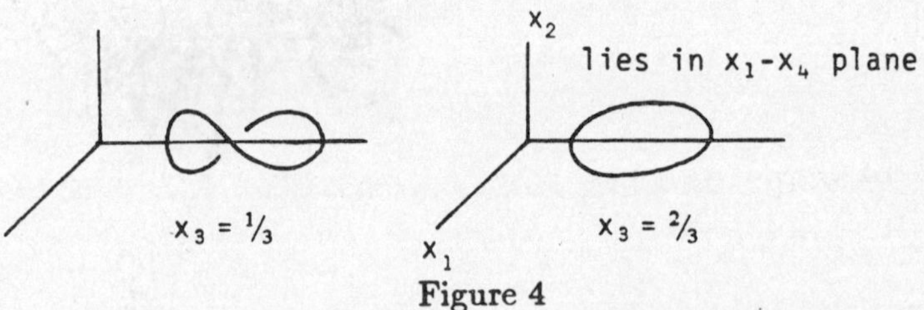

Figure 4

Next, in R^5 we rotate R^4_+ around the x_1, x_2, x_3-axis to obtain an immersed $S^2 \times S^1$, but as we rotate, we spin S^2 in the x_1, x_2 plane by π and thus glue $S^2 \times 0$ to $S^2 \times 2\pi$ by the involution τ (note that $\{(S^2 \times [0, 2\pi])/(x, 0) \sim (\tau(x), 2\pi)\} = S^2 \times S^1$). Then we have a circle $x_4^2 + x_5^2 = 1$ of double points in R^5 and the preimage in $S^2 \times S^1$ is a circle.

The S^1, which is equal to the north pole cross S^1, has a trivial normal bundle both in $S^2 \times S^1$ and R^5, so we may surger it to obtain S^3 and surger the image, using a disk that (north pole) $\times S^1$ bounds in R^5, to obtain an immersion $S^3 \to R^5$.

This gives an immersion $S^3 \times S^1 \to R^5 \times S^1 \subset R^6$ which has a double point set a torus double covering its image. If one wishes, again a circle may be surgered to give an immersion $S^4 \to R^6$ with the same double point data.

Proof of Theorem 1 (B): As in the proof of (A), we can arrange that M^4 is spin bordant to a simply connected spin 4-manifold which we still call M. Then M smoothly immerses in R^6 and by the oriented case bounds an oriented 5-manifold W.

Suppose that the obstruction to extending the spin structure on M over W, $w_2 \in H^2(W, M; Z/2)$ is dual to an orientable 3-manifold N^3 with a trivial normal bundle; choose a trivialization $N \times B^2$. Then M's spin structure extends over $W - (N \times \text{int } B^2)$ and restricts to a spin structure on $N \times S^1$. This gives a spin structure on N^3 which by

Theorem VII.3 spin bounds a spin 4-manifold Y^4. Then the spin structure on $N \times S^1$ extends over $Y^4 \times S^1$, so M^4 spin bounds $W - (N \times \text{ int } B^2) \cup Y^4 \times S^1$.

So our task is to find N^3, which we do by a handlebody argument. Choose a handlebody structure for W with only one 0-handle and no 5-handles. If we attach a 1-handle (necessarily oriented) to the 0-handle, we get $S^1 \times B^4$; the boundary $S^1 \times S^3 = S^3 \times S^1$ can also be achieved by adding a 3-handle to the 0-handle. So we replace all 1-handles by 3-handles. In a similar way, we can replace all 4-handles by 2-handles (since M is connected).

Now we split W between the 2 and 3-handles into W_2 and W_3, where W_2 consists of the 0-handle union the 2-handles and W_3 consists of $M \times I$ union 2-handles (the 3-handles turned upside down), and $\partial W_2 = \partial W_3$. Since $\partial(0-\text{handle}) = S^4$ and M^4 are simply connected, the attaching circle of a 2-handle can be isotoped to an unknot in a coordinate chart; there are two framings for the normal bundle of the circle, so the result of adding the 2-handle is to boundary connected sum with $S^2 \times B^3$ (trivial framing) or $S^2 \tilde{\times} B^3$ (twisted framing). This changes S^4 or M^4 by connected sum with either $S^2 \times S^2$ or $S^2 \tilde{\times} S^2$. The attaching circles of different 2-handles cannot link in 4-space, so we see that $W_2 = B^5 \overset{\partial}{\#} r(B^2 \times B^3) \overset{r}{\#} s(B^2 \tilde{\times} B^3)$ and $W_3 = M \times I \overset{\partial}{\#} p(B^2 \times B^3) \overset{\partial}{\times} q(B^2 \tilde{\times} B^3)$; furthermore $\partial W_2 = \# r(S^2 \times S^2) \# s(S^2 \tilde{\times} S^2) = \partial W_3 = M \# p(S^2 \times S^2) \# q(S^2 \times S^2)$. By sliding 2-handles over 2-handles, we can arrange that $s = q = 1$ (as in Corollary I4.6) unless $s = q = 0$ to begin with; the latter case holds when W is already spin, so we ignore this case and assume $s = q = 1$.

An obstruction to putting a spin structure on W_2 is a fiber $F_2 =$ (point $\times B^3$) of the one copy of $B^2 \tilde{\times} B^3$. An obstruction to extending the spin structure on M^4 over W_3 is also a copy F_3 of the fiber of $B^2 \tilde{\times} B^3$; there is really no difference between the two cases since the 2-handles are all added to a coordinate chart in M^4. If $\partial F_2 = \partial F_3$, we would be finished, but ∂F_2 and ∂F_3 may represent different homology classes in $H_2(\overset{r}{\#} S^2 \times S^2 \# S^2 \tilde{\times} S^2; Z)$.

Since $[\partial F_2]$ and $[\partial F_3]$ are both characteristic classes and have self-intersection zero, it follows from **[Wall4]** that there is an orthogonal (preserves the intersection form) automorphism φ of $H_2(\overset{r}{\#} S^2 \times S^2 \# S^2 \tilde{\times} S^2; Z)$ which takes $[\partial F_2]$ to $[\partial F_3]$. In this case, the automorphism is easy to construct, for by the classification of odd, indefinite forms (if $r \geq 1$), we can choose a basis $[\partial F_2], x_1, x_2, x_3, \ldots, x_{2r+1}$ for $H_2(\partial W_2; Z)$ with intersection form $\begin{pmatrix} 0 & 1 \\ 1 & 1 \end{pmatrix} \overset{r}{\oplus} \langle 1 \rangle \overset{r}{\oplus} \langle -1 \rangle$, and another basis $[\partial F_3], y_1, \ldots, y_{2r+1}$ with the same form. Then the automorphism φ sends $[\partial F_2]$ to $[\partial F_3]$ and x_2 to y_2, $i = 1, \ldots, 2r+1$.

By another theorem of **[Wall1]**, proved later as Theorem 2 in Chapter X, there is a diffeomorphism $g : \partial W_2 \to \partial W_2$ with $g_* = \varphi$. Then we cut W open along ∂W_2, twist ∂W_2 by g^{-1}, and reglue. In the resulting 5-manifold, which we still call W, ∂F_2 and ∂F_3 now represent the same homology class.

It may happen that ∂F_2 and ∂F_3 are not disjoint, but $[\partial F_2] \cap [\partial F_3] = 0$ so the intersection $\partial F_2 \cap \partial F_3$ consists of pairs of points, p_c and q_c, $i = 1, \ldots, n$, of opposite sign. Then we may change ∂F_2, say, by a bordism B consisting of 1-handles, to a surface S of genus n which misses ∂F_3; the construction is exactly like the elimination of triple points in the proof of Lemma 5.

Now, observing that $S \sharp \partial F_3$ is an oriented, connected surface representing $0 \in H_2(\partial W_2; Z)$, we apply Theorem 3 and obtain an orientable 3-manifold Q^3 in ∂W_2 in $\partial Q^3 = S \cup \partial F_3$. The normal bundle to Q^3 in W^5 is trivial since it has one section normal to ∂W_2 and another in ∂W_2 but normal to Q^3. We now add the bordism B to Q^3, pushing $B-(Q \cap B)$ into W_2 so as not to intersect $Q-(Q \cap B)$; note that the normal bundle to $Q \cup B$ is also trivial for the same reason. Now, we add F_3 to one component of $\partial(Q \cup B)$ and a shrunken F_2 to the other component of $\partial(Q \cup B)$ to obtain N^3. The trivialization of the normal bundle extends over F_2 and F_3 since they are 3-balls. □

Much of this argument will be used in Wall's theorem about h-cobordisms (Chapter X) and in Rohlin's theorem (Chapter XI).

IX. $p_1(M) = 3\sigma(M)$, $\Omega_4^{so} = Z$ AND $\Omega_4^{spin} = Z$

The Hirzebruch index theorem [**Hirze**] in dimension 4 is:

THEOREM 1. *If M^4 is smooth, closed and oriented, then the first Pontrjagin number $p_1(M)$ is three times the index of M, $p_1(M) = 3\sigma(M)$.*

COROLLARY 2. *$\Omega_4^{so} = Z$ and the isomorphism is given by the index.*

PROOF: From Theorem VIII.1(b) and the index theorem, we know that if $\sigma(M) = 0$, then M bounds an oriented 5-manifold. But $\sigma(CP^2) = 1$ so $\Omega_4^{so} \xrightarrow{\sigma} Z$ is an isomorphism. □

COROLLARY 3. *$\Omega_4^{spin} \xrightarrow{\sigma/8} Z$ is a monomorphism onto either Z or $2Z$.*

PROOF: As above, if $\sigma(M) = 0$, then M spin bounds. But the index of a spin manifold is divisible by 8. We show below in the proof of the index theorem that the index of the Kummer surface is 16. Until Rohlin's theorem (Chapter XI) is proved we do not know that $\sigma/8$ must be even. □

PROOF OF THEOREM 1: This is normally proved easily by verifying the equality for CP^2 and using the facts that CP^2 generates Ω_4^{so}, and p_1 and σ are bordism invariants and are additive under connected sum. In the order that we are presenting material, we need to use the index theorem to show that $\Omega_4^{so} = Z$, so we give a longer proof, independent of $\Omega_4^{so} = Z$, using $\Omega_4^{spin} = Z$ and a verification of the theorem for the Kummer surface as well as CP^2.

To begin, recall that for a complex surface $p_1 = c_1^2 - 2c_2$ (see [**M-S**], page 177); also recall that the total Chern class of CP^n is $(1+\alpha)^{n+1} = 1 + (n+1)\alpha + \dots$ where α generates $H^2(CP^n; Z)$ and is dual to CP^{n-1}, and α^n generates $H^{2n}(CP^n; Z) = Z$. Thus $c_1(CP^2) = 3\alpha$ and $c_2(CP^2) = 3\alpha^2$. Then $p_1(CP^2) = c_1^2(CP^2) - 2c_2(CP_2) = (3\alpha)^2 - 6\alpha^2 = 3\alpha^2$. Thus $p_1(CP^2) = 3 = 3\sigma(CP^2)$.

Next, we need to verify the index theorem for spin 4-manifolds. At this stage we do not know the generator of $\Omega_4^{spin} = Z$, but the Kummer surface K^4 represents some multiple m of the generator M^4. So $p_1(K) = mp_1(M) = m\ell\sigma(M) = \ell\sigma(K)$ for some integer ℓ. If we check that $p_1(K) = -48$ and $\sigma(K) = -16$, then we will know that $\ell = 3$ and $p_1 = 3\sigma$ for spin manifolds.

We need a construction of the Kummer surface from which it is easy to calculate both p_1 and σ. One can get p_1 (via $c_1^2 - 2c_2$) by describing K^4 as a non-singular quartic in CP^2, but the index is not easily gotten from this description; vice versa, one may easily calculate σ from a handlebody description of K, but not p_1. So we proceed from our earlier description of $CP^2 \overset{9}{\#} (-CP^2)$ with its torus T^2 representing the obstruction to a spin structure (see Chapter V). This T^2 has a trivial normal bundle $T^2 \times B^2$ and

$T^2 \times (B^2 - 0)$ inherits the Lie group spin structure from $CP^2 \overset{9}{\#} (CP^2) - T^2$. We will form a 5-dimensional bordism W^5 by taking two copies of $CP^2 \overset{9}{\#} (-CP^2) \times I$ and attaching $T^2 \times B^2 \times [-1,1]$ by gluing $T^2 \times B^2 \times -1$ to one copy along $T^2 \times B^2 \times 1$ and $T^2 \times B^2 \times 1$ to the other copy along $T^2 \times B^2 \times 1$. To extend the orientations on the two copies of $CP^2 \overset{9}{\#} (-CP^2)$ over the "handle", we must glue by the "identity" on one end and by the identity composed with a reflection in the B^2-factor on the other end. Then the Lie group spin structure induced on $T^2 \times S^1 \times \{-1 \cup 1\}$ extends trivially.

Thus we have a bordism from $2(CP^2 \overset{9}{\#} (-CP^2))$ to a spin 4-manifold K^4 (which is actually a Kummer surface). Then $p_1(K^4) = p_1(2(CP^2 \overset{9}{\#} (-CP^2))) = -48 = 3(-16) = 3\sigma(2(CP^2 \overset{9}{\#} (-CP^2))) = 3\sigma(K^4)$, and we have the index theorem for spin 4-manifolds.

Now let M^4 be an arbitrary smooth, closed, oriented 4-manifold. It is bordant to a simply connected manifold and the obstruction to finding a spin structure can be taken to be a torus T^2 with self-intersection $T \cdot T = n$. Then the pairwise connected sum $(M^4, T^2) \overset{n}{\#} (-CP^2, CP^1)$ will still have a torus T^2 as the obstruction to a spin structure on $M^4 \overset{n}{\#} (-CP^2) = M'$ and now $T \cdot T = 0$. So T has a trivial normal bundle $T^2 \times B^2$ whose boundary $T^2 \times S^1$ either inherits the Lie group spin structure or inherits a spin structure which does not coincide on a circle C in T^2 with the Lie group structure. In the latter case we can glue a $B^2 \times S^1 \times B^2$ to $(M^4 \overset{n}{\#} (-CP^2)) \times I$ along $C \times S^1 \times B^2 \times 1$ and obtain a bordism from $M^4 \overset{n}{\#} (-CP^2)$ to a spin manifold. In the case of the Lie group structure on T^3, we glue $(CP \overset{9}{\#} -CP^2) - (T^2 \times \text{ int } B^2)) \times [-1,1]$ to $(M^4 \overset{n}{\#} (-CP^2)) \times I$ along $T^3 \times [-1,1] \times 1$ to obtain a bordism from $M^4 \overset{n}{\#} (-CP^2)$ to a spin manifold disjoint union $CP^2 \overset{9}{\#} (-CP^2)$. Since we know the index theorem for spin manifolds and for $\pm CP^2$, then we know it for M^4 and hence all oriented, closed, smooth 4-manifolds. □

Note that we have shown that M^4 is bordant to the union of a spin manifold and a connected sum of CP^2's and $-CP^2$'s. One way to attempt to show that $\Omega_4^{so} = Z$ (with the isomorphism given by the index) is to show that a spin 4-manifold connected sum with some $\pm CP^2$'s is diffeomorphic to a connected sum of $\pm CP^2$'s; this is fairly easy to do for the Kummer surface (see [**H-K-K**], §2, pg. 67) which, in fact, is the generator of $\Omega_4^{spin} = Z$ (see Chapter XI). But until Rohlin's theorem is proved using $\Omega^{spin} = Z$ in Chapter XI, we do not know whether there is a strange spin manifold of index 8.

PROPOSITION 4. *Let M^4 be closed, smooth and simply connected. The homotopy type of M determines its tangent bundle, T_M.*

PROOF: M is homotopy equivalent to a wedge of 2-spheres with a 4-cell attached. Over the wedge, T_M is determined by the second Stiefel–Whitney class $w_2(T_M)$, a homotopy invariant. Over the 4-cell T_M is determined by the Euler class $\chi(M)$ and the first Pontrjagin number $p_1(M)$ which is equal (via the index theorem) to $3\sigma(M)$, and both $\chi(M)$ and $\sigma(M)$ are obvious invariants of homotopy type. □

X. WALL'S DIFFEOMORPHISMS AND H-COBORDISM

In 1964 Wall published [**Wall1,2**] some basic theorems about realizing automorphisms of the intersection form by diffeomorphisms and that manifolds with isomorphic forms are h-cobordant.

THEOREM 1. *Let M_0^4 and M_1^4 be smooth, simply connected, closed 4-manifolds with isomorphic intersection forms. Then M_0 and M_1 are h-cobordant.*

(The theorem also holds in the topological case if the Kirby–Siebenmann triangulation invariants are equal [**Quinn1**], [**Freedman1**]. Moreover, the h-cobordism is a product in the topological case [**Freedman1**], but not necessarily in the smooth case [**Donaldson3**].)

THEOREM 2. *Let N^4 be a smooth, simply connected, closed 4-manifold with an intersection form which is either indefinite or has rank ≤ 8. Then any automorphism of the intersection form of $M = N \,\sharp\, S^2 \times S^2$ is realized by a diffeomorphism of M.*

PROOF OF THEOREM 1: First we do the spin case and then make some comments on the changes necessary for the non-spin case. Since M_0 and M_1 have the same index, $M_1 \cup (-M_0)$ bounds a connected spin manifold W^5. Choose a Morse function $f : W \to [0,1]$ with $f^{-1}(0) = M_0$, $f^{-1}(1) = M_1$, and consider the associated handlebody structure on W. We can always (in any dimension) cancel the 0-handles and similarly the 5-handles.

Let R^4 be a chart in M_0. A 1-handle $B^1 \times B^4$ in W is attached to M_0 (actually, a thickened M_0), and we may assume by connectivity of M that the attaching $S^0 \times B^4$ lies in R^4. After adding the 1-handle, M_0 becomes $M_0 \,\sharp\, S^1 \times S^3$, to which all other handles are attached. We could also achieve $M_0 \,\sharp\, S^1 \times S^3$ by adding a 3-handle $B^3 \times B^2$ along a trivial $S^2 \times B^2$ in R_0^4, so we trade this 1-handle for the 3-handle and then attach all other handles in the same way as before. This changes W^5 by a surgery on the S^1 determined by the 1-handle; we still call the result W^5 and note that it is still spin. In this way we change all 1-handles to 3-handles and all 4-handles to 2-handles (by inverting W so that k-handles become $(5-k)$-handles).

So W is built by adding 2- and 3-handles to M_0. Since M_0 is simply connected, each attaching circle of a 2-handle can be isotoped to a trivial circle in R_0^4. The framing is zero in $\pi_1(SO(3)) = Z/2$ because W is spin. Thus the result of adding the 2-handles, say k of them, to M_0^4 is to obtain a bordism to $M_0 \overset{k}{\sharp} S^2 \times S^2$, and similarly with M_1^4. If we assume that all 2-handles correspond to critical points at which our Morse function is $< 1/2$, and for 3-handles, $f > 1/2$, then we have shown that $f^{-1}(1/2) = M_{1/2} = M_0 \overset{k}{\sharp} S^2 \times S^2 = M_1 \overset{k}{\sharp} S^2 \times S^2$.

This is worth being called:

THEOREM 3. *If M_0 and M_1 are smooth, closed, 1-connected 4-manifolds with isomorphic forms, then $M_0 \overset{k}{\#} S^2 \times S^2$ is diffeomorphic to $M_1 \overset{k}{\#} S^2 \times S^2$.*

(We have only proved this in the spin case, but the non-spin case follows because for odd M_0, $M_0 \# S^2 \tilde{\times} S^2$ is diffeomorphic to $M_0 \# S^2 \times S^2$ as in Corollary I.4.6 and Remark.)

Continuing with the proof of Theorem 1, we want to cut W along $M_{1/2}$ and reglue a different way so that the result is an h-cobordism. More precisely, let

$$d : M_0 \overset{k}{\#} S^2 \times S^2 \to M_1 \overset{k}{\#} S^2 \times S^2$$

be the diffeomorphism, and let ϕ be the isomorphism between the intersection form for M_0 and that for M_1, that is, $\phi : H_2(M_0; Z) \to H_2(M_1; Z)$ preserves the intersection form. We would like each 2-handle to be cancelled homologically by a 3-handle. Given a 2-handle $B^2 \times B^3$ let the corresponding $S^2 \times S^2$ have coordinates so that $S^2 \times 0$ comes from the core of the 2-handle $B^2 \times 0$, and $0 \times S^2$ is $0 \times \partial B^3$. For the 3-handle $B^3 \times B^2$, let $S^2 \times 0$ be $\partial B^3 \times 0$ and $0 \times S^2$ come from $0 \times B^2$. Then we want an isomorphism $\psi : H_2(\overset{k}{\#} S^2 \times S^2; Z) \to H_2(\overset{k}{\#} S^2 \times S^2; Z)$ to consist of $\overset{k}{\oplus} \begin{pmatrix} 1 & 0 \\ 0 & 1 \end{pmatrix}$.

It follows that we need a diffeomorphism g of $M_1 \overset{k}{\#} S^2 \times S^2$ to itself which on homology satisfies $(gd)_* = g_* d_* = \phi \oplus \psi$. Then if we glue W back together by the diffeomorphism gd, the 2- and 3-handles will cancel homologically so that we have a homology-cobordism; everything is simply connected so it is an h-cobordism.

To find g, we need to realize the automorphism $g_* = (\phi \oplus \psi) d_*^{-1}$ of $H_2(M_1 \overset{k}{\#} S^2 \times S^2)$ by a diffeomorphism g. This can be done by Theorem 2 since $N = M_1 \overset{k}{\#} S^2 \times S^2$ has indefinite form, and we can gain another $S^2 \times S^2$ by adding a (geometrically) cancelling pair of 2- and 3-handles to W by a "birth".

In the case when M_1^4 is odd, the bordism W will not be spin, there may be 2-handles attached to M_0 with the non-trivial framing, and the middle level of W, $M_{1/2}$ is $M_i \overset{j}{\#} S^2 \times S^2 \overset{\ell}{\#} S^2 \tilde{\times} S^2$. But from Corollary I.4.6 and its proof, we see that it can be arranged for all 2-handles of W to be attached by the trivial framing and $M_{1/2}$ is $M_i \overset{k}{\#} S^2 \times S^2$. Then the proof proceeds as in the even case. □

PROOF OF THEOREM 2: We will show, via the handle calculus, how to realize a certain type of automorphism by a diffeomorphism of M, and then quote Wall for the fact that these automorphisms generate all automorphisms.

First assume that M has a handlebody decomposition with no 1- or 3-handles. Then M has a framed link L such as in Figure 1 and an associated intersection matrix (using as basis the 2-dimensional classes determined by the 2-handles) as in Figure 2a.

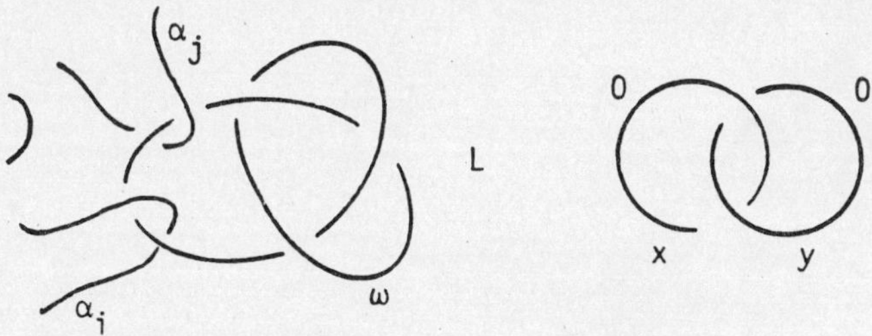

Figure 1

$$\begin{pmatrix} \alpha_1\cdot\alpha_1 & \dots & \alpha_1\cdot\omega & 0 & 0 \\ \vdots & & & & \vdots \\ & & & & \\ \alpha_1\cdot\omega & & \omega\cdot\omega & 0 & 0 \\ 0 & & 0 & 0 & 1 \\ 0 & \dots & 0 & 1 & 0 \end{pmatrix} \begin{matrix} \alpha_1 \\ \vdots \\ \alpha_k \\ \omega \\ x \\ y \end{matrix} \qquad \begin{pmatrix} \alpha_1\cdot\alpha_1 & \dots & \alpha_1\cdot\omega & \alpha_1\cdot\omega & 0 \\ \vdots & & & & \vdots \\ & & & & \\ \alpha_1\cdot\omega & & \omega\cdot\omega & \omega\cdot\omega & 0 \\ \alpha_1\cdot\omega & & \omega\cdot\omega & \omega\cdot\omega & 1 \\ 0 & \dots & 0 & 1 & 0 \end{pmatrix} \begin{matrix} \alpha_1 \\ \vdots \\ \alpha_k \\ \omega \\ x+\omega \\ y \end{matrix}$$

a $\qquad$ b

$$\begin{pmatrix} \alpha_1\cdot\alpha_1 & \dots & \alpha_1\cdot\omega & 0 & 0 \\ \vdots & & & & \vdots \\ & & & & \\ \alpha_1\cdot\omega & & \omega\cdot\omega & 0 & 0 \\ 0 & & 0 & \omega\cdot\omega & 1 \\ 0 & \dots & 0 & 1 & 0 \end{pmatrix} \begin{matrix} \alpha_1-(\alpha_1\cdot\omega)\cdot y \\ \vdots \\ \alpha_k-(\alpha_k\cdot\omega)\cdot y \\ \omega-(\omega\cdot\omega)\cdot y \\ x+\omega \\ y \end{matrix}$$

c

Figure 2

The plan is to slide handles over handles to get a new framed link L'; this does not exactly give a diffeomorphism of M, but rather a diffeomorphism from the 4-manifold M_L to the 4-manifold M'_L. However, if after sliding handles we return to exactly the original framed link L, then we will have an auto-diffeomorphism from $M = M_L$ to itself. We will do this and see what happens to H_2 by keeping track of the basis.

First, slide the component x over the component denoted by w obtaining L' in Figure 3. w is an arbitrary indivisible element of $H_2(N;Z)$ and therefore can be taken to be a basis element. This changes the basis by replacing x by $x+w$ and the intersection form changes to the matrix in Figure 2b. Now we slide various handles over y, aiming to restore zeros to the next-to-the-last row and column.

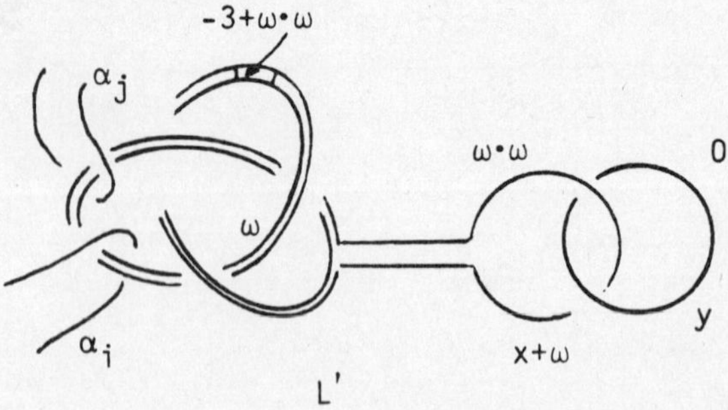

Figure 3

As in Lemmas I.4.4, I.4.5, we move y around $x+w$ removing the linking of each α_i with $x+w$ and then removing the linking of w with $x+w$. This requires (algebraically) $\alpha_1 \cdot w$ slides for $\alpha_1, \alpha_2 \cdot w$ slides for $\alpha_2, \ldots$ and $w \cdot w$ slides for w. We obtain the matrix and basis in Figure 2c, and the framed link L'' in Figure 4.

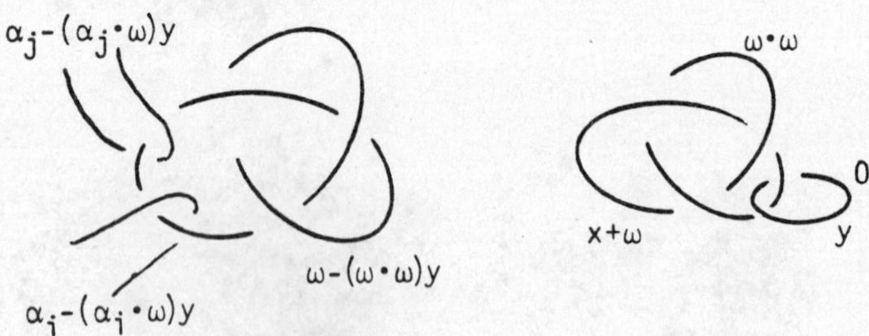

Figure 4

Now we slide $x+w$ over y so as to unknot $x+w$ and so as to reduce the framing from $w \cdot w$ to 0 or 1. In the even (spin) case, we slide $x+w$ over y algebraically $(w \cdot w)/2$ times and get framing zero on $x+w-1/2(w \cdot w)y$, and we have obtained exactly L again; the basis has changed according to the automorphism A_w where $A_w(\alpha_i) = \alpha_i(\alpha_i \cdot w)y$, $i = 1, \ldots, k$, $A_w(w) = w - (w \cdot w)y$, $A_w(x) = x + w - 1/2(w \cdot w)y$, and $A_w(y) = y$. In the case where $w \cdot w$ is odd, we have achieved a diffeomorphism from $N \sharp S^2 \times S^2$ to $N \sharp S^2 \tilde{\times} S^2$ where $S^2 \tilde{\times} S^2$ is denoted by $w \cdot w$ (link diagram) 0 and A_w is as before except that $A_w(x) = x + w$.

In the even case, we are done because the automorphisms A_w, and the automorphisms A'_w constructed with the roles of x and y reversed, and the automorphisms induced by diffeomorphisms of $S^2 \times S^2$ connect sum the identity on N, generate the orthogonal group of the intersection form on $M = N \sharp S^2 \times S^2$, (see [**Wall1**], page 136, and [**Wall2,4**]). In the odd case we only need the observation that $N \sharp S^2 \times S^2 = N \sharp S^2 \tilde{\times} S^2$

(Corollary I.4.6) and the existence of the diffeomorphism of $S^2 \tilde{\times} S^2 = CP^2 \sharp -CP^2$ which is complex conjugation on CP^2.

The case when N^4 has 1- and 3-handles can be taken care of by arranging through handle slides for each 1-handle to be cancelled by a 2-handle, each 3-handle to be cancelled by a 2-handle, and the remaining 2-handles $\alpha_1, \ldots, \alpha_k, w$ to have algebraically zero intersection with the cocores of the 1-handles and the cores of the 3-handles; in other words, the boundary maps at the chain level from the 3-handles to the 2-handles, and the 2-handles to the 1-handles, should be of the form $\begin{pmatrix} I & 0 \\ 0 & 0 \end{pmatrix}$. Then the argument proceeds as before with $\alpha_1, \ldots, \alpha_k, w, x, y$. $\square$

XI. ROHLIN'S THEOREM

§1. An elementary proof.

We begin with an elementary proof of Rohlin's theorem: If M^4 is closed, smooth and spin, then index$(M) \equiv 0(16)$.

COROLLARY. *$\Omega_4^{\text{spin}} \xrightarrow{\sigma/16} Z$ is an isomorphism, and the Kummer surface is a generator.*

For a French translation of Rohlin's original proof, as well as Y. Matsumoto's geometric proof, see **[G-M]**. The proof given here is more along the lines of **[F-K]** except that the technical difficulties are fewer because spin structures are used.

PROOF: By the usual method we can kill $\pi_1(M)$ by adding 2-handles to $M \times I$ along the generating circles of $\pi_1(M)$ with the correct framing so that the spin structure on M extends across the bordism. The new boundary, still called M, is now simply connected.

Assume, contrary to the theorem, that index$(M) \equiv 8(16)$. By connected summing with copies of the Kummer surface with the right orientation, we can assume that index$(M) = 8$. Now let $N^4 = M \# CP^2 \overset{9}{\#} -CP^2$. Then index$(N) = 0$ and an integral dual to $\omega_2(N)$ is the homology class $[T^2] \in H_2(N; Z)$ where T^2 is the imbedded torus defined in Chapter V.

Since $[T^2]\cdot[T^2] = 0$, T^2 has a trivial normal circle bundle $\sum_{T^2}$ which is diffeomorphic to T^3. As shown in Chapter V, $N^4 - T^2$ is spin, and (uniquely, since $H^1(N^4 - T^2; Z) = 0$) T^3 inherits a spin structure which is the Lie group spin structure. Any section of $\sum_{T^2}$ gives T^2 the Lie group spin structure (see below), and we want to use this fact, that our T^2_{Lie} is non-zero in Ω_2^{spin}, to show that N cannot exist and therefore index $M \equiv 8(16)$ is impossible.

Since index$(N^4) = 0$, N bounds an orientable 5-manifold, W^5, which we can assume has no 1 or 4 or 5-handles (see the proof of Theorem X.1). The middle level, between the 2 and 3-handles, is obtained from S^4 by surgering the unlink and therefore is $\overset{r}{\#} S^2 \times S^2 \overset{s}{\#} S^2 \tilde{\times} S^2$ (by Corollary I.4.6 we can take $s = 1$). The middle level is also obtained from N^4 by surgering the unlink (since $\pi_1(N) = 0$) and therefore is $N^4 \overset{p}{\#} S^2 \times S^2$ (use Corollary I.4.6 to eliminate any $S^2 \tilde{\times} S^2$).

A characteristic surface for $N^4 \overset{p}{\#} S^2 \times S^2$ is still T^2. Under the diffeomorphism g between $N^4 \overset{p}{\#} S^2 \times S^2$ and $\overset{r}{\#} S^2 \times S^2 \# S^2 \tilde{\times} S^2$, $g_*[T^2]$ is a characteristic class in $H_2(\overset{r}{\#} S^2 \times S^2 \# S^2 \tilde{\times} S^2; Z)$. Now $g_*[T^2]$ is represented by a smooth imbedded 2-sphere; this can be constructed by hand or can be deduced as follows from Wall's theorems.

The fiber, S^2, of $S^2 \tilde{\times} S^2$ is a characteristic surface for $\overset{r}{\#} S^2 \times S^2 \# S^2 \tilde{\times} S^2$. There is an orthogonal automorphism of $H_2(\overset{r}{\#} S^2 \times S^2 \# S^2 \tilde{\times} S^2; Z)$ which carries $g_*[T^2]$ to $[S^2]$

since both have self-intersection zero. This is a special case of [**Wall3**, Theorem 4], but can easily be seen by applying the method of proof of Theorem II.3.2 to $g_*[T^2]$; (since $g_*[T^2]\cdot g_*[T^2]=0$ and is characteristic, it must have a dual of square 1 and together these define a form $\begin{pmatrix}0&1\\1&1\end{pmatrix}$ whose orthogonal complement must be $\overset{r}{\oplus}\begin{pmatrix}0&1\\1&0\end{pmatrix}$, so we have the desired automorphism).

This automorphism can be realized by a diffeomorphism h, according to Theorem X.2, perhaps after connected summing with another copy of $S^2\times S^2$. Thus $hg(T^2)$ and S^2 represent the same homology class in $H_2(\overset{r}{\#}\, S^2\times S^2 \,\#\, S^2\tilde{\times} S^2; Z)$. According to Theorem II.1.1 there is a bordism Y^3 from $hg(T^2)$ to S^2 which is smooth and oriented and lies in $(\overset{r}{\#}\, S^2\times S^2\,\#\, S^2\tilde{\times} S^2)\times I$. Furthermore, Y^3 is an integral dual to ω_2, so its complement has a (unique) spin structure. (This follows from the proof of Theorem II.1.1 since Y^3 is the inverse image of CP^1 which is an integral dual to $\omega_2(CP^2)$.)

The normal circle bundle C to Y in W is oriented (by choosing the outward pointing normal vector to C), so by Proposition IV.2, it gets a spin structure from $W-Y$. There is a non-zero section of C over a neighborhood of a 1-skeleton of Y. Choose a basis of circles for $H_1(Y;Z/2)$ and lift a neighborhood of these circles into C by the section; these neighborhoods are oriented by Y, so by Proposition IV.2 again, they get a spin structure from C. Since there exists some spin structure on Y, it follows from Proposition IV.1 that the section and C determine a spin structure on Y.

On ∂Y, this spin structure must be the unique one on S^2 and must be a spin structure on T^2 gotten from a section of its normal circle bundle T^3_{Lie}. But any T^2 in T^3_{Lie} gets the Lie group spin structure. So Y^3 is a spin bordism from T^2_{Lie} to S^2, different elements of Ω_2^{spin}, and we have contradicted the assumption that index $M\equiv 8(16)$. □

§2. Ω_4^{char}.

Although the above proof is the most elementary proof of Rohlin's Theorem that we know, the theorem has a more natural setting. The material below follows [**F-K**], but is easier because our homomorphism $\phi:\Omega_4^{\text{char}}\to Z/2$ is easier to define and show well defined.

Let (M^4,F^2) be a characteristic pair, where M^4 is smooth, closed and oriented, F^2 is closed, oriented, and dual to $\omega_2(M^4)$, and M^4-F^2 is assumed to have a given spin structure which does not extend past F^2. Let Ω_4^{char} be bordism of characteristic pairs; that is, (M_1,F_1) is bordant to (M_2,F_2) if there exists a pair (W^5,Y^3) with W and Y smooth and oriented, Y dual to $\omega_2(W)$, and a spin structure on W^5-Y^3 (not extending past Y^3), such that $\partial(W,Y)=(M_1,F_1)-(M_2,F_2)$ and spin structures agree.

LEMMA 1 [**F-K**]. $\Omega_4^{\text{char}} = Z\oplus Z$ *and an isomorphism is given by* $(\sigma(M)$, $(F\cdot F-\sigma(M))/8)$. *The generators are* (CP^2,CP^1) *and* $(CP^2\,\#\,-CP^2, 3CP^1\,\#\,CP^1)$.

We sketch the proof in [**F-K**]. The homomorphism is clearly onto, so we must show that if $\sigma(M)=0=F\cdot F$, then (M,F) bounds. As in the proof of Rohlin's theorem, bord M^4 to a 1-connected 4-manifold, preserving F, and let W^5 be a smooth 5-manifold with $\partial W=M^4$ and no 1, 4 or 5-handles. Then, as before in Chapter X, Theorem 1, W splits between the 2 and 3-handles into two pieces W_2 and W_3, where $W_2=$

$\overset{r}{\#} S^2 \times B^3 \overset{\varepsilon}{\#} S^2 \tilde{\times} B^3$, $\varepsilon = 0$ or 1, and is obtained from the 0 and 2-handles, and $W_3 = M^4 \times I \overset{s}{\#} S^2 \times B^3$ is obtained from the 3-handles, that is, $M^4 \times I$ union 2-handles which are dual to the 3-handles. Note that $\partial W_3 = M^4 \cup (-\partial W_2)$, and $F \times I$ is characteristic in W_3 and $\varepsilon(* \times B^3)$ is characteristic in W_2. Since $F \times 1$ and $\varepsilon(* \times \partial B^3)$ have self-intersection zero and are characteristic in ∂W_2, (possibly $[F] = 0$ and $\varepsilon = 0$), there exists a diffeomorphism $d : \partial W_2 \to \partial W_2$ so that $d_*[F \times 1] = [\varepsilon(* \times \partial B^2)]$. Then cut W^5 along ∂W_2 and reglue via d obtaining W_d. Then since $d_*[F]$ is the same homology class as $[\varepsilon(* \times \partial B^3)]$ there is a connected 3-manifold U in a collar of ∂W_2 joining $F \times 1$ to $\varepsilon(* \times \partial B^3)$. Then $F \times I \cup U \cup \varepsilon(* \times B^3)$ is a characteristic 3-manifold Y in W_d. It is not hard to check that the spin structure on $M^4 - F$ extends to $W_d - Y$, but not across Y. □

Next, we define a homomorphism $\phi : \Omega_4^{\text{char}} \to Z/2$. Let (M^4, F^2) be a characteristic pair. The normal circle bundle C to F in M gets a spin structure from $M - F$. Pick a basis of circles for $H_1(F; Z/2)$ and, using any section of C, push a neighborhood of the circles into C. The spin structure on C puts a spin structure on this neighborhood and hence, via Proposition IV.1, a spin structure on F. That this spin structure on F is independent of the choice of section is due to the fact that F is characteristic and therefore a normal circle has the Lie group spin structure. For, different sections will differ by multiples of the normal circle, but the spin structure on C corresponds to an element σ of $H^1(E; Z/2)$, where E is the principal $SO(3)$-bundle of T_C, and σ is zero on any normal circle to F.

Thus F gets a well defined spin structure which determines an element in $\Omega_2^{\text{spin}} = Z/2$. To see that it is independent of the choice of characteristic pair (M, F), apply the same argument to a bordism (W^5, Y^3) to put a spin structure on Y^3 which gives the spin bordism between two choices of characteristic pairs.

THEOREM 2. *$\phi : \Omega_4^{\text{char}} \to Z/2$ coincides with $\theta : \Omega_4^{\text{char}} \to Z/2$ defined by*

$$\theta(M, F) = \frac{F \cdot F - \sigma(M)}{8} \bmod 2.$$

Proof: It suffices to check this on generators of $\Omega_4^{\text{char}} = Z \oplus Z$. Clearly, $\theta(CP^2, CP^1) = 0 = \phi(CP^2, CP^1)$ since $CP^1 = S^2$. Also $\theta(CP^2 \# -CP^2, 3CP^1 \# CP^1) = 1$. ϕ is also 1 because $3CP^1 \# CP^1$ is represented by T^2 and $T^2 \cdot T^2 = 8$. After connected sum with eight copies of $(-CP^2, CP^1)$, we have our old friend from Chapter V, namely a T^2 whose circle bundle is T^3_{Lie}. □

Rohlin's Theorem is now an easy corollary; if M is spin, take F^2 to be empty, so $0 = \phi(M, -) = \theta(M, -) = -\sigma(M)/8$ (2), so $\sigma(M) \equiv 0(16)$.

COROLLARY 3 [K-M]. *If the dual to w_2 in M_4 can be represented by a smooth, imbedded 2-sphere, then $(F \cdot F - \sigma(M))/8 \equiv 0(2)$.*

§3. The μ-invariant and the Arf invariant of a knot.

We can now define a $Z/2$ invariant, Rohlin's invariant, of a Z-homology 3-sphere Σ^3. Let $\rho(\Sigma^3) = \sigma(M^4)/8 \bmod 2$ where M^4 is a spin 4-manifold with spin boundary Σ^3 (note that Σ^3 has only one spin structure, and that the intersection form on M is even and unimodular, hence σ is divisible by 8). A different choice M' gives the same value for $\rho(\Sigma^3)$ since $\sigma(M) - \sigma(M') = \sigma(M \cup -M') = 0 \bmod 16$.

Now suppose that Σ^3 is only a $Z/2$-homology 3-sphere, so that it still has a unique spin structure but no longer will $\sigma(M^4)$ be divisible by 8. Now we define a $Z/16$ invariant $\mu(\Sigma^3) = \sigma(M^4) \bmod 16$ and note that if Σ^3 is also a Z-homology 3-sphere then $\mu(\Sigma^3)$ is 0 or 8.

Finally, an arbitrary oriented 3-manifold with a specified spin structure can be given this same $Z/16$ invariant $\sigma(M^4) \bmod 16$.

EXAMPLES: The Poincaré homology 3-sphere bounds the E_8 form and hence its Rohlin invariant is $1 \in Z/2$. The lens space $L(p, p-1)$ defined by p-surgery on the unknot bounds (for p odd) the even 4-manifold determined by either of the framed links in Figure 1, whose

Figure 1

index is $-(p-1)$ and μ-invariant is $(1-p) \bmod 16$. (The first framed link may be obtained by blowing up $p-1$ (-1)-circles to reduce the framing from p to 1, see Figure 2a, and then blowing down the 1-circle, or by blowing up a succession of -1-circles which reduce the p framing and successively split off the previous -1-circles, as in Figure 2b, and finally blowing down the 1-circle.)

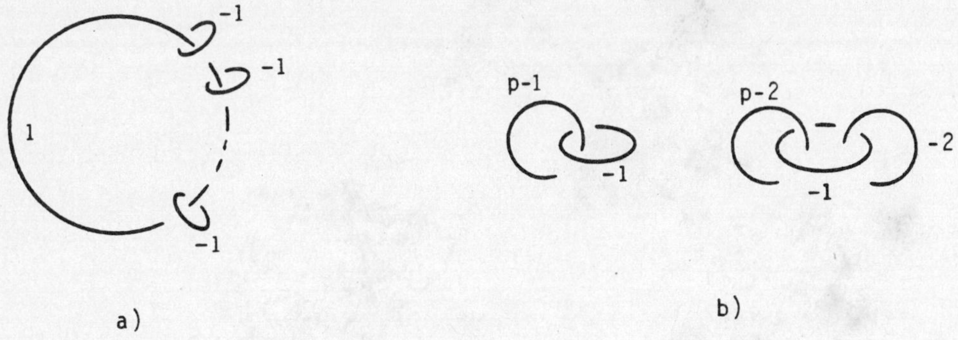

Figure 2

We can also define the Arf invariant of a knot K in S^3 (or, in fact, in any Z-homology 3-sphere) using spin structures and Theorem 2. Let F^2 be a Seifert surface for K in $S^3 \times 1$ and push the interior of F into the interior of $S^3 \times [-1,1]$. There is a unique spin structure on $(S^3 \times [-1,1]) - F^2$ which does not extend across F; it restricts to the unique spin structure on $S^3 \times -1$, and to the unique one on $(S^3 \times 1) - K$ which does not extend across K. As in the construction in the proof of Theorem 2, we get a spin structure on the normal circle bundle to F^2, and then get a spin structure on F and hence on the closed surface $\hat{F} = F \cup B^2$. This determines an element of $\Omega_2^{\text{spin}} = Z/2$ which we define as $A(K)$.

We show that $A(K)$ is independent of the choice of Seifert surface F by use of Theorem 2: first add a 2-handle to $S^3 \times [-1,1]$ along K with framing 1 (any odd framing would work). The new manifold has F union the cone of the 2-handle, i.e. $\hat{F}$, as characteristic surface, and we can cap off $S^3 \times -1$ with B^4 and the new boundary with a spin 4-manifold to obtain M^4. Clearly $A(K) = \phi(M, \hat{F})$ which is independent of the choice of F since any two Seifert surfaces are bordant in $S^3 \times I$.

COROLLARY 4. *$A(K)$ equals the Rohlin invariant of the homology 3-sphere obtained by $+1$ surgery on K. (If K is a knot in an arbitrary Z-homology 3-sphere Σ, then this equality must be corrected by $\rho(\Sigma)$.)*

The Arf invariant of a knot K, Arf(K), is normally defined as follows: for a Seifert surface F for K, $H_1(F; Z/2)$ is a $Z/2$ vector space with non-singular inner product given by the intersection form. We define a quadratic form $q : H_1(F; Z/2) \to Z/2$ by $q(\gamma)$ equals the number of full twists, mod 2, in a neighborhood of an imbedded circle representing $\gamma \in H_1(F; Z/2)$, as in Figure 3. It is easy to verify that $q(\gamma_1 + \gamma_2) = q(\gamma_1) + q(\gamma_2) + \gamma_1 \cdot \gamma_2$ so there is an Arf invariant in $Z/2$ associated with $(H_1(F; Z/2), q)$, (see the Appendix) and this is Arf(K). A formula is given by $\text{Arf}(K) = \sum_{i=1}^{g} q(\gamma_{2i-1}) \cdot q(\gamma_{2i}) \bmod 2$ where $\gamma_1, \ldots, \gamma_{2g}$ is a standard hyperbolic basis for $H_1(F^2; Z/2)$.

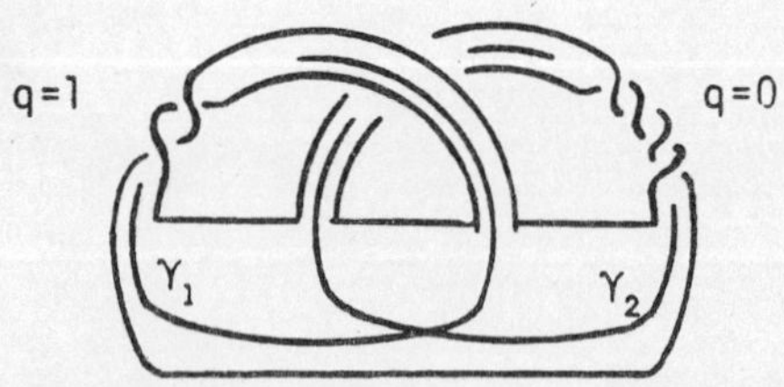

Figure 3

Of course one can directly show that Arf(K) is independent of the choice of F, but this will follow from showing that Arf(K) = $A(K)$ (see **[Rober]**). To compute $A(K)$, we need to determine the spin structure on $T_F \mid \gamma_i$. Choose the Lie group framing on $T_F \mid \gamma_i$, that is, the framing given by the tangent vector τ to γ_i and a normal vector n_1 to γ_i in T_F. Follow (τ, n_1) by the normal vector n_2 to F in S^3, and then add the inward pointing normal ν to S^3 in B^4. Then the trivialization (τ, n_1, n_2, ν) on $T_{B^4} \mid \gamma_i$ either agrees or disagrees with the unique trivialization of T_{B^4}, according to whether (τ, n_1, n_2) agrees or disagrees with the unique trivialization of T_{S^3} over the 2-skeleton. (τ, n_1, n_2) agrees iff (n_1, n_2) rotates an odd number of times compared with the 0-framing of γ_i (this is obvious if γ is an unknot). Thus if (n_1, n_2) rotates an odd number of times, then γ_i gets the Lie group spin structure, but if (n_1, n_2) rotates evenly, then (τ, n_1) was the wrong trivialization and we should have given $T_F \mid \gamma_i$ the other trivialization, that is, the one that bounds, or is zero in Ω_1^{spin}.

Therefore F gets a spin structure which is determined by the rule: for even (odd) full twists in F around γ_i, the spin structure on $T_F \mid \gamma_i$ bounds (does not bound).

Finally, we note that if we define a quadratic form $q' : H_1(F; Z/2) \to Z/2$ by

$$q'(\gamma_i) = \begin{cases} 0 & \text{if } \gamma_i \text{ has a bounding spin structure} \\ 1 & \text{if } \gamma_i \text{ has a non-bounding spin structure} \end{cases}$$

then q' is quadratic and identical to q so their Arf invariants are equal. But it is easy to see that the Arf invariant of q' gives an isomorphism $\Omega_2^{\text{spin}} \to Z/2$. We have shown

LEMMA 5 [**Rober**]. *For a smooth knot K in S^3 (or a homology 3-sphere),* $\text{Arf}(K) = A(K)$.

COROLLARY 6. *Let M^4 be a closed 4-manifold and F^2 a characteristic surface which is smoothly imbedded in M^4 except for n singular points at which the imbedding is locally pairwise homeomorphic to the cone on (S^3, K_i) for knots K_i, $i = 1, \ldots, n$, in S^3. Then there is a congruence extending Theorem 2:*

$$\frac{K \cdot K - \sigma(M)}{8} + \sum_{i=1}^{n} \text{Arf}(K_i) \equiv \phi(K) \pmod 2.$$

The reader can also devise a formula for the case when F^2 is a smoothly immersed surface with n double points. In a 4-ball neighborhood of a double point, F^2 looks like the cone on the Hopf link. If we replace the two disks by an annulus inside the 4-ball, then we have added a torus to F^2; the generator corresponding to the cone of the annulus has one twist in it, so the quadratic form is one, whereas the other generator corresponds to an arc leaving the double point along one sheet and returning along the other, so its twists must be computed from the description of F^2.

COROLLARY 7. *Given a spun 3-manifold N^3, an oriented 4-manifold M^4 with $\partial M^4 = N$, and an oriented surface F^2 in M^4 which is dual to the obstruction to extending N's spin structure over M^4, then*

$$\mu(N^3) = \sigma(M^4) - F \cdot F \pmod{16}.$$

If F has singular points as in Corollary 6, then we must correct by $8\sum_{i=1}^{n} \text{Arf}(K_i)$. The choice of $\sigma(M^4) - F \cdot F$ rather than $F \cdot F - \sigma(M^4)$, a choice of sign, is dictated by the example of $L(3,1)$. It occurs as the link of a singularity in a complex curve and when the singularity is resolved, we see that $L(3,1)$ bounds the unknot with framing -3, so that $\mu(L(3,1)) = -1 - (-3) = 2$. Via the sequence in Figure 4, we see that $L(3,1)$ also bounds a spin 4-manifold with index 2.

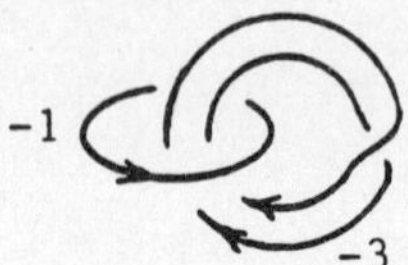

Figure 4

We end this chapter with a few more computations of invariants.

First we return to the lens space $L(p, p-1)$ defined by surgery on the unknot with framing p, which is the boundary of B^4 with a 2-handle attached to the unknot with framing p, called M^4. Then the μ-invariant is also equal mod 16 to

$$\sigma(M) - F \cdot F = 1 - p$$

where F^2 is the characteristic surface consisting of the cone of the 2-handle union cone on the unknot. More generally, the homology lens space obtained by p-surgery on a knot K has μ-invariant equal mod 16 to

$$\sigma(M) - F \cdot F + 8 \text{ Arf}(K) = 1 - p + 8 \text{ Arf}(K).$$

Still more generally, we compute the μ-invariant of a $Z/2$-homology sphere obtained from surgery on a framed link L by finding a characteristic sublink (the sums of the elements in this sublink should link an even element evenly and an odd element oddly), band connected summing the circles in the sublink to form a knot K and computing its framing f, whereupon the μ-invariant is given by

$$\text{index(linking matrix)} - f + 8 \text{ Arf}(K) \text{ mod } 16.$$

This can be seen to be equivalent to adding 2-handles to the framed link L, finding a characteristic surface with one singular point equal to the cone on K, and then using our earlier formulae. For example, in Figure 5, the sum of the components is characteristic and represented by a trefoil knot with framing zero, so the μ-invariant is $0 - 0 + 8 = 8$ which is correct since the 3-manifold is the Poincaré homology sphere.

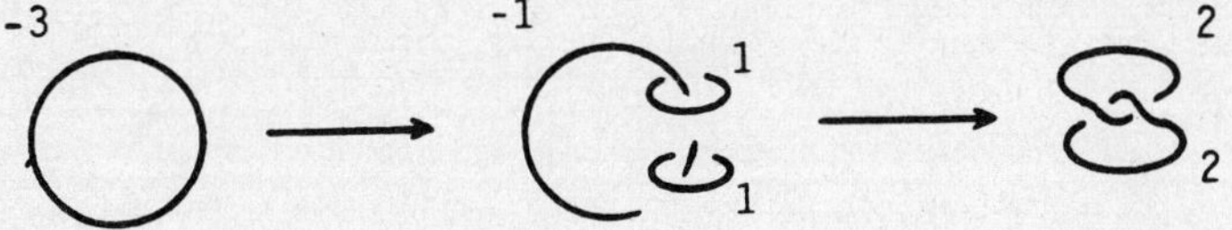

Figure 5

XII. CASSON HANDLES

§1. Whitney's Trick.

To motivate Freedman's work, we must describe Casson's work; to motivate Casson's work we must see how a higher dimensional topologist would try to classify simply connected 4-manifolds, what goes wrong, and how Casson made progress.

Suppose we want to represent the bilinear form $-(E_8 \oplus (1))$ by a closed, smooth 4-manifold (remember that this is impossible (III, §3)), knowing that $-(E_8 \oplus (1)) \oplus \begin{pmatrix} 0 & 1 \\ 1 & 0 \end{pmatrix}$ is represented by $M^4 = CP^2 \overset{10}{\#} (-CP^2)$.

Let $\alpha, \beta \in H_2(M;Z)$ generate the hyperbolic pair, i.e., $\alpha \cdot \alpha = \beta \cdot \beta = 0$, $\alpha \cdot \beta = 1$, and $\alpha \cdot \gamma = \beta \cdot \gamma = 0$ for all other basis elements γ. Since $\pi_1(M) = 0$, α and β are represented by maps $f : S^2 \to M$ and $g : S^2 \to M$. We can assume that f and g are smooth immersions so that $f(S^2)$ and $g(S^2)$ have transverse intersections and self-intersections. The algebraic sum of the intersections between $f(S^2)$ and $g(S^2)$ is one since $\alpha \cdot \beta = 1$. We can arrange that the

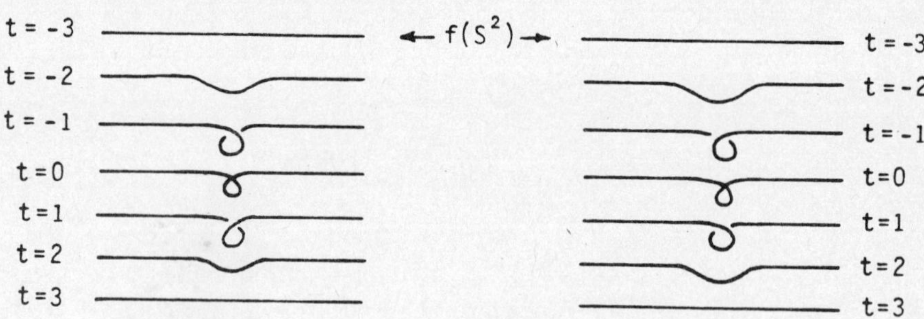

Figure 1.1 *In a coordinate chart with coordinates (x,y,z,t) we see 3-dimensional slices as t ranges from -3 to $+3$. In these slices we see slices of $f(S^2)$ with a new double point when $t = 0$. They have opposite signs.*

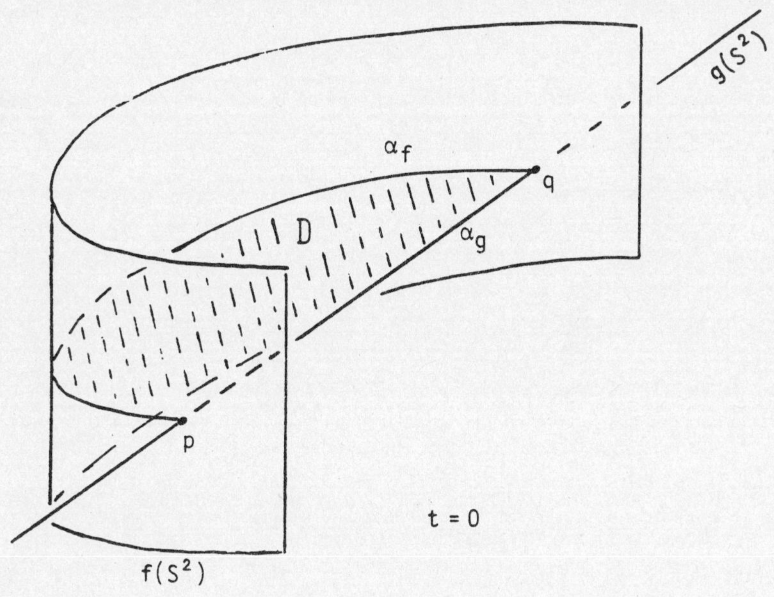

Figure 1.2

algebraic sum of the self-intersections of both $f(S^2)$ and $g(S^2)$ is zero by introducing more self intersections with the correct sign. This can be done locally, as in Figure 1.1.

Whitney's trick [**Whitney**] is a method for removing a pair of double points of opposite sign; it works in dimension ≥ 5. To understand Whitney's trick, imagine the following (Figure 1.2) model in dimension 4; p and q are the two points of intersection (of opposite sign) of, say, $f(S^2)$ and $g(S^2)$ in M^4; α_f and α_g are curves in $f(S^2)$ and $g(S^2)$ respectively joining p and q. Our picture is of $R^3 = [(x,y,z,t) \in F^4 \mid t = 0\}$ and we suppose that a neighborhood of α_f in $f(S^2)$ lies in R^3 as drawn, and that a neighborhood of α_g on $g(S^2)$ is of the form (α_g, t), $t \in (-1,1)$, so that the only part of this neighborhood that we see in time $t = 0$ is α_g. Furthermore, suppose, as drawn, that there is a 2-ball D (the Whitney disk) with $\partial D = \alpha_f \cup \alpha_g$ and that D is smoothly imbedded with $D \cap f(S^2) = \alpha_f$, $D \cap g(S^2) = \alpha_g$, and normal bundle $D \times R^1 \times R^1$ where $f(S^2) \cap (D \times R^1 \times R^1) = \alpha_f \times R^1 \times 0$ and $g(S^2) \cap (D \times R^1 \times R^1) = \alpha_g \times 0 \times R^1$.

In this model it is easy to describe how to remove the pair of double points—simply push $f(S^2)$ across D by an isotopy which has support in the normal bundle of a slightly larger disk D' (see Figure 1.3).

The model is basically the same in higher dimensions; the dimensions of $f(S^2)$ and $g(S^2)$ may be larger, but p, g, α_f, α_g and D are the same, and trivial normal bundles of D must split in two factors, one for α_f and one for α_g just as in our case. Then the isotopy in higher dimensions is the same as ours except that it is the identity on the

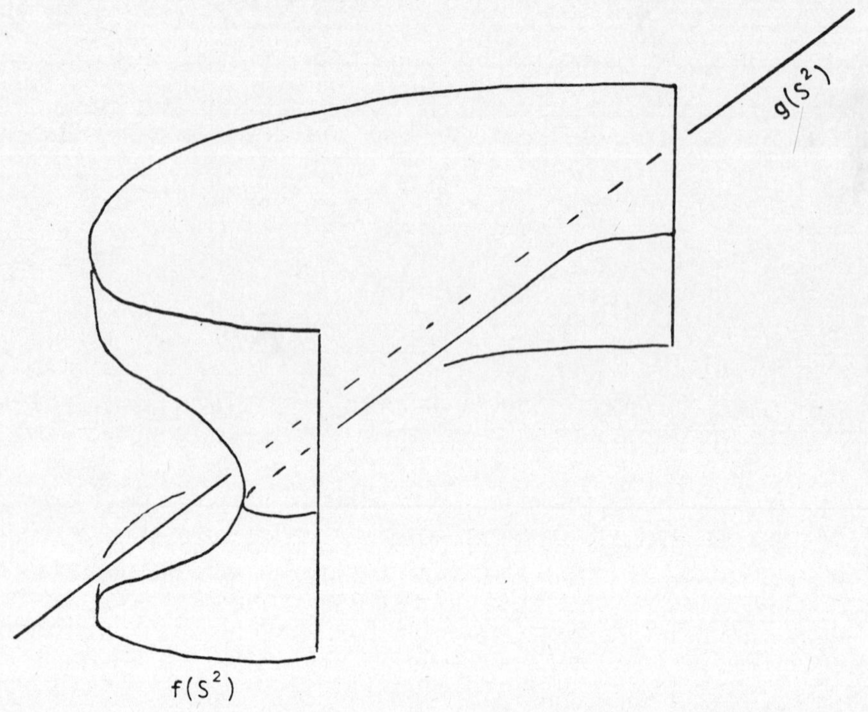

Figure 1.3

extra dimensions.

To carry out the Whitney trick, it is necessary to be able to construct the model. α_f and α_g always exist (assuming connectivity) and usually $\alpha_f \cup \alpha_g$ is null homotopic so that D can be immersed. In all higher dimensions, this immersion can be a smooth imbedding via transversality, but in dimension 4, it is no easier to imbed D than to imbed the 2-spheres in the first place! It is this difficulty only that makes dimension 4 so much harder than higher dimensions—in fact, Freedman proves that D can be imbedded topologically and then things proceed fairly easily. There can be one further problem, to make the normal bundle of D match up correctly with $f(S^2)$ and $g(S^2)$, but this can be analyzed and dealt with.

Because of Rohlin's Theorem, it is known that Whitney's trick must fail in certain smooth cases (see [**F-K**]).

§2. Finger Moves.

Without Whitney's trick, progress in 4-manifolds was slow. It was Casson's brilliant idea in 1973-74 to avoid the problem of imbedding 2-spheres and instead find a substitute for a 2-sphere that was easier to imbed. Casson lectured on those ideas in April 1974 at I.H.E.S. and they appear in [**G-M**].

Consider again our favorite 4-manifold M^4 with form $-(E_8 \oplus 1) \oplus \begin{pmatrix} 0 & 1 \\ 1 & 0 \end{pmatrix}$ (columns labelled α, β). We want to smoothly imbed 2-spheres representing α and β and meeting at one point, i.e. $S^2 \vee S^2$. Their normal disk bundles would give $S^2 \times S^2$-(open 4-ball); we will describe this 4-manifold as a 0-handle (4-ball) with two 2-handles attached to the Hopf link with 0-framing. It is trivial to imbed the 0-handle in M, but the 2-handles may have intersections and self-intersections. We cannot avoid the self-intersections, but we can avoid the intersections between the two 2-handles (let's call them h^2_α and h^2_β) as follows.

First immerse h^2_α. If the attaching circle for h^2_β is null homotopic in the complement of (0-handle union h^2_α), then via transversality we can move the null homotopy to an immersion giving an h^2_β disjoint from h^2_α. Perhaps Casson's most surprising idea was that by increasing the number of double points of h^2_α via "finger moves", he could arrange that β was null homotopic. So we digress to explain finger moves, which will be used over and over in the construction of Casson handles.

Let K^2 be a 2-dimensional complex (e.g., an immersed 2-sphere) in a 4-manifold (e.g., M^4). Suppose we change K^2 to a complex L^2 by a regular homotopy as drawn in Figure 2.1, i.e., we place our finger tip on a 2-simplex in K^2, push around an element α of $\pi_1(M - K)$, and push back through the 2-simplex creating two new double points. The push through K^2 is exactly the same as the reverse of the Whitney trick, and there is a 2-ball D which would serve as the Whitney disk (see §1).

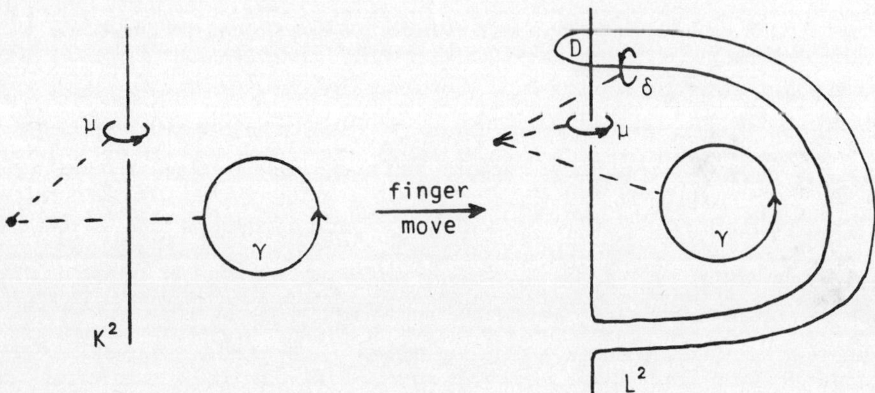

Figure 2.1

What happens to $\pi_1(M - K^2)$ as a result of the finger move? The finger move kills the commutator $[\mu, \gamma^{-1}\mu\gamma]$. To see this, notice that $\pi_1(M - K)$ is not changed by the isotopy in Figure 2.2(a), nor by adding an arc as in Figure 2.2(b) (since the arc is codimension 3, it does not effect the fundamental group). The complement of $L \cup D$ can be deformed into the complement of ($K \cup$(arc)), so that $\pi_1(M - K) = \pi_1(M - (L \cup D))$. It remains to see that adding D to $M - (L \cup D)$ kills $[\mu, \gamma^{-1}\mu\gamma]$.

A neighborhood of P is pairwise homeomorphic to $(R^4, R^2_{xy} \cup R^2_{zt})$ where R^2_{xy} is

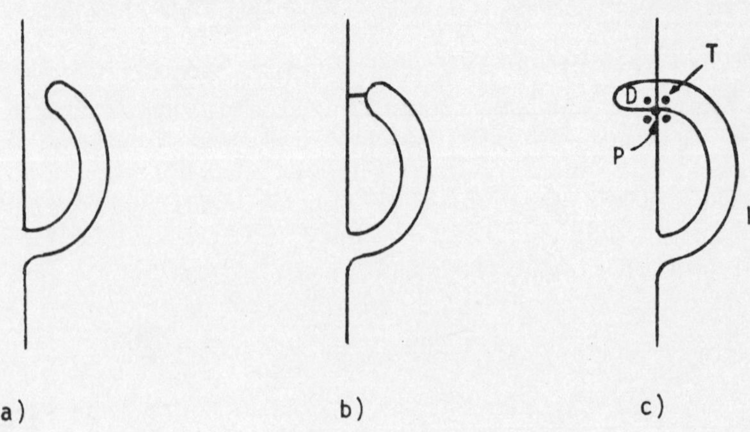

Figure 2.2

the $x-y$-plane and similarly R^2_{zt}. A torus $T^2 = S^1 \times S^1$ links $R^2_{xy} \cup R^2_{zt}$ in R^4 (in fact, $T^2 = \{(x,y,z,t) \in R^1 \mid x^2+y^2 = 1 = z^2+t^2\}$). This torus T^2, called the *distinguished* torus, intersects D in one point, say q. Then the normal circle ν to D at q is a commutator $[\mu,\delta] = \mu\delta\mu^{-1}\delta^{-1}$ where μ and δ generate the torus (see Figure 2.1). Removing D from $L \cup D$, or equivalently, adding D to $M-(L \cup D)$ allows ν to bound the normal disk to D at q, so that $\nu \simeq 1$ in $M-L$ and we have killed $[\mu,\delta]$. Finally, observe that $\delta = \gamma^{-1}\mu\gamma$ if one keeps track of base points, so we have killed $[\mu,\gamma^{-1}\mu\gamma]$.

§3. Casson Handles.

Now we go back to M^4 and the immersed 2-handle h^2_α which is attached to the circle α in $\overset{\alpha}{\underset{0}{}}\bigcirc\!\!\bigcirc\overset{\beta}{\underset{0}{}}$ in $S^3 = \partial$(0-handle). For simplicity, call an immersed 2-handle a *kinky* handle and let $M_0 = M^4 -$ (0-handle). It is not hard to check that $\pi_1(M_0 - h_\alpha)$ is generated by conjugates of the circle β since β is isotopic to a normal circle to h_α. (This fact is analogous to $\pi_1(S^3$-knot) being generated by finitely many conjugates of a normal circle to the knot.) Furthermore, $\pi_1(M_0 - h_\alpha)$ is a perfect group, since $H_1(M - h_\alpha; Z) = 0$ because the circle β generates $H_1(M_0, \partial M_0; Z)$ (algebraically the only point of intersection between α and β in $H_2(M;Z)$ is the 0-handle). Thus any element in $\pi_1(M_0 - h_\alpha)$ is a product of commutators of conjugates of β, but by a finite number of finger moves, we can kill these commutators of conjugates. Thus we get a kinky handle h_α with $\pi_1(M_0 - h_\alpha) = 0$ so that we then get a kinky handle h_β disjoint from h_α. Again use finger moves in $M_0 - h_\alpha$ to ensure that $\pi_1(M_0 - h_\alpha - h_\beta) = 0$.

Of what use are these immersions? Here is Casson's surprising program. Suppose for simplicity that h_α has only one double point p and let C be a circle in h_α through p which leaves p along one place and returns along the other (see Figure 3.1). C is homotopically trivial in $M_0 - (h_\alpha \cup h_\beta)$ so C bounds a kinky handle $h_{\alpha,1}$ meeting $h_\alpha \cup h_\beta$ only in C. (Technically, we have to be careful whether C is on the core 2-ball of h_α or on the boundary of the 4-dimensional immersed handle h_α; the choices are

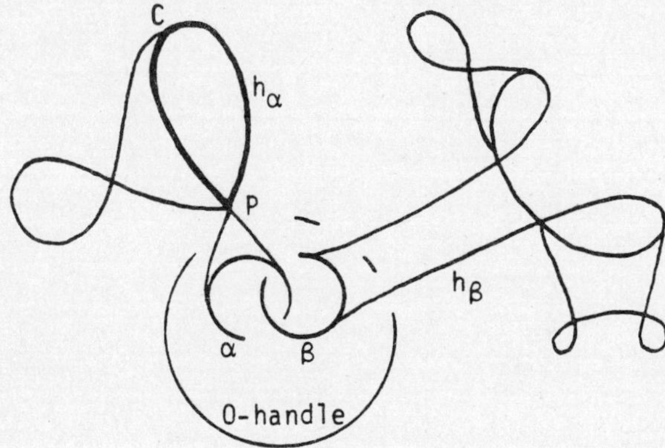

Figure 3.1

isotopic via an imbedded annulus and we will hereafter ignore this point.)

If $h_{\alpha,1}$ was imbedded, then $h_\alpha \cup h_{\alpha,1}$ would be diffeomorphic to an imbedded 2-handle and we would be done. Presumably, $h_{\alpha,1}$ is not imbedded, but we still use this fact to determine the framing of C to which $h_{\alpha,1}$ is attached; use the framing so that $h_\alpha \cup h_{\alpha,1}$ has a chance to be imbedded 2-handle (this happens to be framing zero as in Figure 3.7 later).

In the same way, we add kinky handles $h_{\beta,1}$ and $h_{\beta,2}$ to h_β as in Figure 3.1. Using more finger moves when necessary (it is neçessary to show that only the last stage of immersed 2-handles need to be altered by finger moves), we iterate this process. Add third stage kinky handles $h_{\alpha,i,j}$ and $h_{\beta,i,j}$, and then fourth stage kinky handles, and so on. The end result after countable iterations is two Casson handles (see Figure 3.2 for the simplest—only one kink at each stage—Casson handle); each is the union of countably many kinky handles with all boundary deleted except for the attaching part, namely $S^1 \times$ int B^2.

There is an uncountable collection $\mathcal{W}$ of Casson handles, uncountable because at each stage the kinky handles can have any finite number of kinks (self-intersection points). Our discussion so far motivates Casson's main theorem:

THEOREM 3.1 (Casson) **[G-M]**, **[Edwards2]**. *Let M^4 be simply connected and let $D_1, \ldots, D_n$ be smoothly immersed transverse 2-disks in M^4 with boundaries, $\partial D_1, \ldots, \partial D_n$, imbedded disjointly in ∂M^4, and $D_i \cdot D_j = 0$ for $i \neq j$. Assume that there exist $\beta_1, \ldots, \beta_n \in H_2(M; Z)$ such that $\beta_i \cdot \beta_i$ is even and $D_i \cdot \beta_j = \delta_{ij}$. Then the D_i's can be regularly homotoped* (rel ∂) *to be disjoint, and then kinky handles may be added disjointly so as to build n disjoint, smoothly imbedded Casson handles with $D_1, \ldots, D_n$ as their first stages. Furthermore, these Casson handles satisfy*

Figure 3.2 Simplest Casson handle

Property 1: *Each Casson handle is proper homotopy equivalent*, rel $\partial = S^1 \times \text{ int } B^2$, *to* $B^2 \times \text{ int } B^2$.

Property 2: *Each Casson handle is a smooth submanifold of* $B^2 \times B^2$ *with* $S^1 \times \text{ int } B^2 = \partial B^2 \times \text{ int } B^2$.

Freedman proved:

Property 3: *Each Casson handle is homeomorphic to* $B^2 \times \text{ int } B^2$, rel ∂.

In our case with $M_0 = CP^2 \overset{10}{\#} (-CP^2) - \text{ int } B^4$, h_α and h_β correspond to D_1 and D_2, and β serves as a dual for h_α and α serves as a dual for h_β, that is, $h_\alpha \cdot h_\beta = 0$, $h_\alpha \cdot \beta = 1$, $h_\beta \cdot \alpha = 1$, and $\alpha \cdot \alpha = \beta \cdot \beta = 0$. So we have sketched a proof of the first part of Theorem 3.1 in our special case.

To understand Property 1, note that any Casson handle, CH, is simply connected. For $\pi_1(h_\alpha)$ was generated by C, but C was killed by $h_{\alpha,1}$. The generators of $\pi_1(h_\alpha \cup h_{\alpha,1})$ are killed by the third stage kinky handles, and so on (see Figure 3.7). It is not hard to check that the homology of CH is that of a 2-handle, but to do so we have to be precise about the boundary of CH. A Casson handle is attached to a thickened circle, an $S^1 \times B^2$, which can be seen already in the first kinky handle h_α. All other boundary including $\partial(S^1 \times B^2)$ is assumed deleted from the Casson handle, for we want it to be pairwise proper homotopy equivalent (in fact, homeomorphic) to $(B^2 \times R^2, S^1 \times R^2)$. There is an alternate description of a Casson handle which is crucial for Property 2 and Freedman's work. To begin we give a handlebody description of the simplest Casson handle; later we relate it to the Whitehead continuum.

Consider first one kinky handle k with one kink (self-intersection). It is not hard to see that the kinky handle k is diffeomorphic to $S^1 \times B^3$—just pick a curve like C above (running from the double point along one sheet back to the double point on the other sheet) and shrink k into a neighborhood of C. But it is important to keep track of the attaching circle of k (the part of ∂k along which k is attached to a 4-manifold is called $\partial_- k$ and equals $S^1 \times B^2$). It is drawn in Figure 3.3.

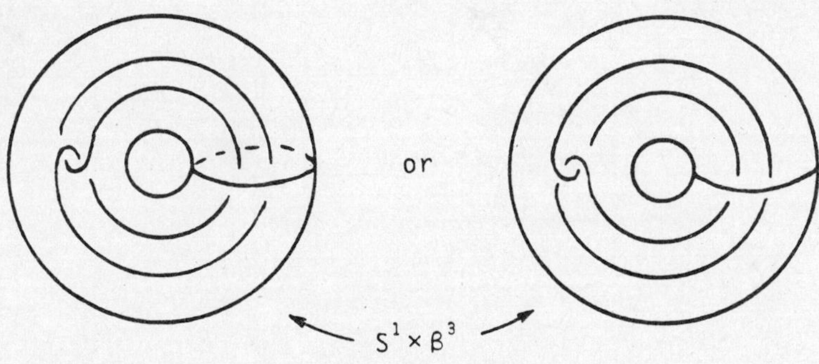

Figure 3.3

To see this, examine a neighborhood of the double point of the form $B^2 \times B^2$ where $B^2 \times 0$ and $0 \times B^2$ are parts of the core of the kinky handle, and note that $\partial(B^2 \times 0)$ and $\partial(0 \times B^2)$ are linked circles in $\partial(B^2 \times B^2) = S^3$ (see Figure 3.4).

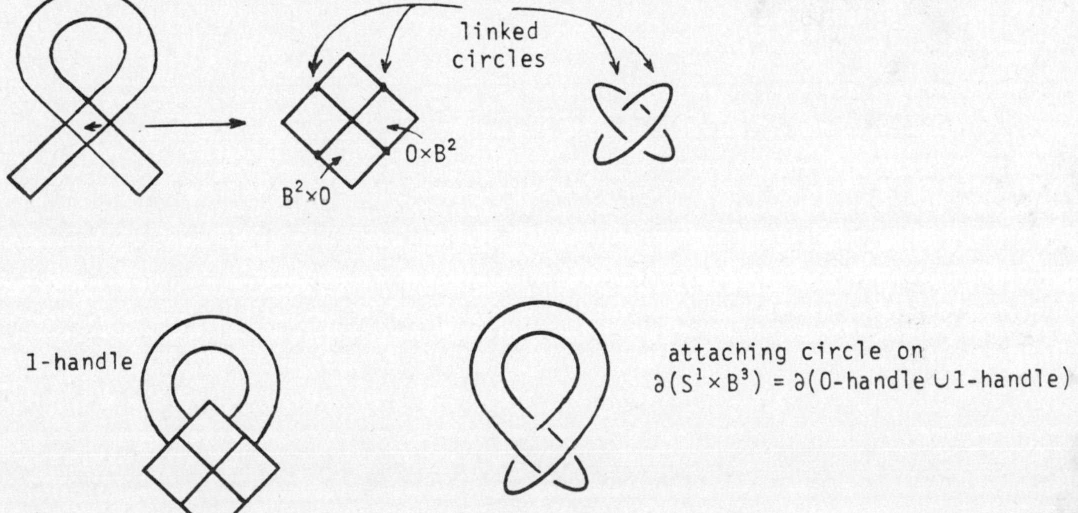

Figure 3.4

To obtain the rest of k we must add a one-handle $B^1 \times B^3$ to the $B^2 \times B^2$ so that the two-dimensional 1-handle $B^1 \times B^1 \times 0 \times 0$ in $B^1 \times B^3$ is attached to the 0-handles $B^2 \times 0$ and $0 \times B^2$ so as to make the core of k (Figure 3.4). Then the position of the attaching circle of k is evident (the two possibilities in Figure 3.3 arise from the choice of how to add the 1-handle).

The most convenient notation for k is in Figure 3.5 where (recall I, §2) a circle with a dot denotes the 1-handle obtained from B^4 by removing the obvious slice disk for the circle.

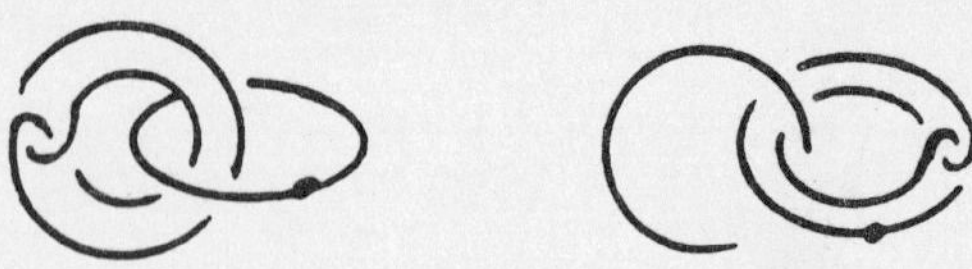

Figure 3.5

When we add a second kinky handle to the first, we attach it so as to kill $\pi_1(k)$, as drawn in Figure 3.6(a). This process is iterated to get the simplest Casson handle in Figure 3.7.

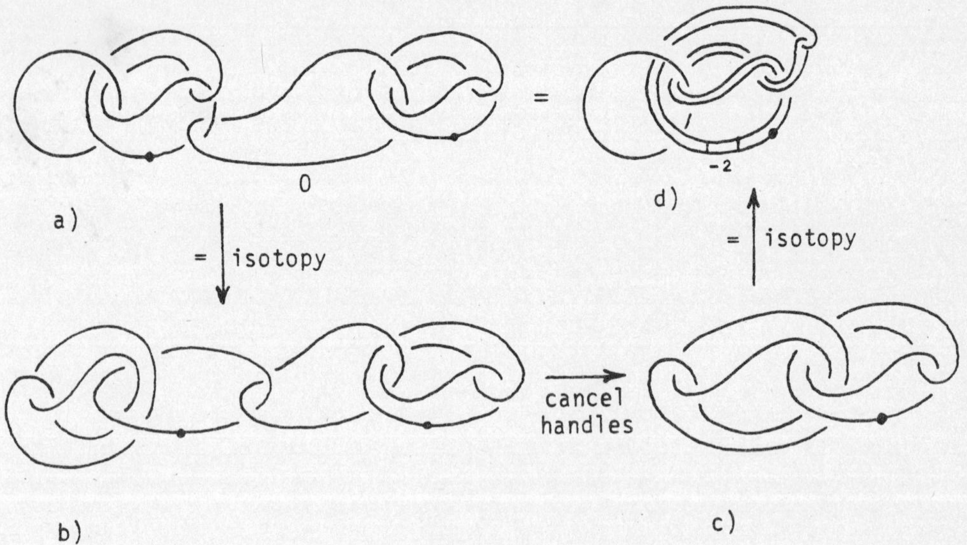

Figure 3.6(a), (b), (c), (d)

Note that if we ignore $\partial_- CH$, the attaching circle of CH (the left-most undecorated circle), then the first pair of 1- and 2-handles cancel (just erase them), as do the next pair, and the next, and so on; thus all the handles cancel and we are left with R^4. Therefore the interior of CH is *smoothly* trivial, and it is only as a pair, $(CH, \partial_- CH)$, that the Casson handle is interesting. Now suppose we cancel, keeping track of the attaching circle, as in Figure 3.6(b). The first pair of handles cancel as in Figure 3.6(c) which "doubles" the attaching circle or doubles the 1-handle (Figure 3.6(d)). Cancelling another pair of handles results in either redoubling the attaching circle or the 1-handle as in Figure 3.8. More cancelling gives more doubling, so we can see that the pair $(CH, \partial_- CH)$ is not at all trivial.

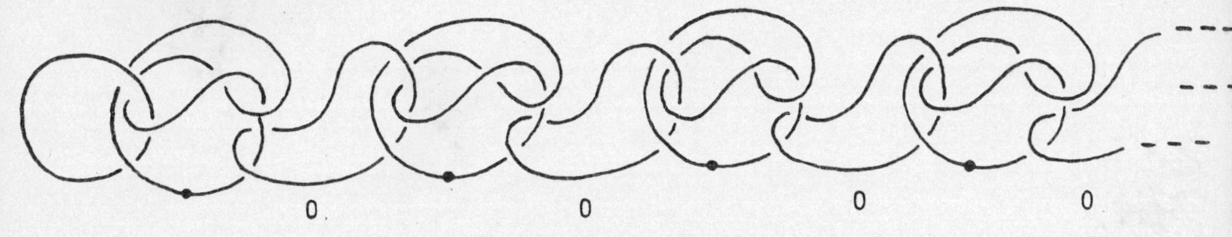

Figure 3.7

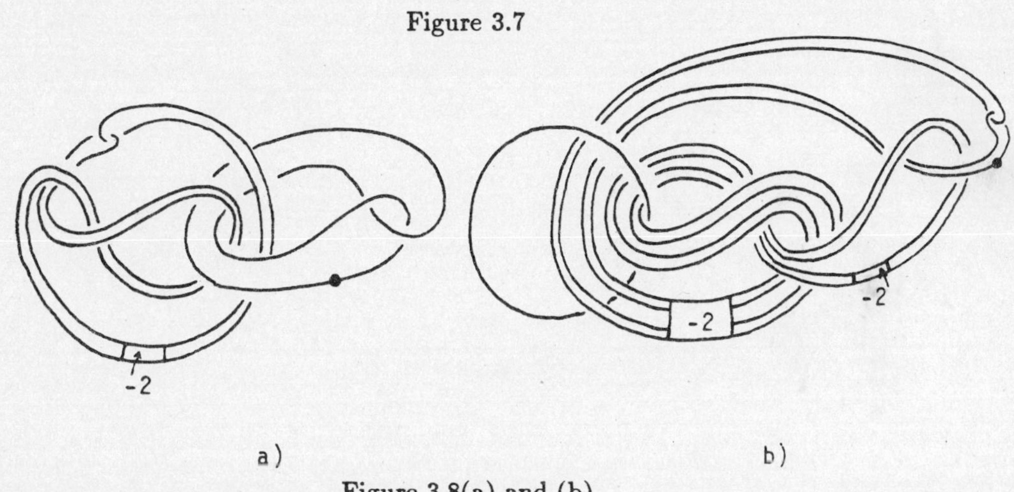

Figure 3.8(a) and (b)

Figures 3.5(b), 3.6(d) and 3.8(b) are particularly interesting. Let T_n be an n-stage tower, i.e., the union of the first n kinky handles (then $T_\infty = CH$). The figures indicate that T_1, T_2 and T_3 are equal to $B^2 \times B^2$ with a certain slice disk removed; moreover, $\partial_- T = S^1 \times B^2$ with attaching circle $S^1 \times 0$ and the dotted circle lies in $B^2 \times S^1$. More is true, for the slice disk for the dotted circle in Figure 3.6(d) can be chosen to lie inside a normal disk bundle neighborhood of the slice for the dotted circle in the previous state, Figure 3.7(b). Similarly the slice in Figure 3.8(b) lies in a thickening of the slice in Figure 3.7(d). This holds for all n, i.e., T_n is $B^2 \times B^2$ minus a slice for a dotted circle in $B^2 \times S^1$ which lies in the previous slice thickened. And $T_\infty = CH$ is $B^2 \times$ int B^2 minus the intersection of the thickened slices! This is related to a well-known space, the Whitehead continuum.

§4. The Whitehead Continuum.

The Whitehead continuum Wh is constructed as follows: Let T_0 be the solid torus $S^1 \times B^2$ and let T_1 be another solid torus smoothly imbedded in T_0, as drawn in Figure 4.1. If we imbed another solid torus in T_1 just as T_1 is imbedded in T_0 then we get T_2, and so on. (The pair (T_k, T_{k+1}) should be diffeomorphic to the pair (T_0, T_1).) The intersection $\bigcap_{k=0}^{\infty} T_k$ is called the Whitehead continuum, Wh.

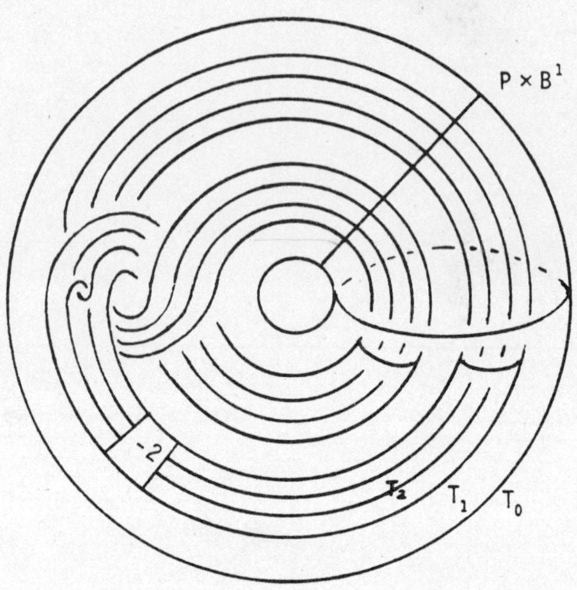

Figure 4.1

In the early thirties, J. H. C. Whitehead gave a proof that any contractible non-compact 3-manifold was homeomorphic to R^3, discovered a mistake, and found [**White**] a counterexample, namely, $S^3 - Wh$. $S^3 - Wh$ is not simply connected at infinity, so S^3/Wh (or R^3/Wh) is not a manifold. However, $(R^3/Wh) \times R$ is R^4. We will discuss these properties of Wh and its relation to Casson handles after a closer look at Wh.

If we delete one point from Wh, namely, the limit point of the clasps , then what remains is a homogeneous space which is locally homeomorphic to the Cartesian product of a Cantor set and the reals. This can be seen by examining the intersection of Wh with $p \times B^2 \subset S^1 \times B^2 = T_0$. Since the T_k, $k = 0, 1, 2, \ldots$, get thinner and may be chosen to be centered on a line, we see that $Wh \cap p \times B^2 = Wh \cap p \times B^1$. Then we also see that T_1 misses the first, third, and fifth fifths of $p \times B^1$ if we divide B^1 into five intervals. Similarly T_2 misses the odd fifths of the remainder, and so on, giving a Cantor set homeomorphic to the usual one where we remove middle thirds (in particular, ours is homeomorphic to the real numbers which contain no ones or threes in their expansions using the integers mod 5, e.g., .242042...).

Using the standard imbedding of T_0 in S^3 or R^3, we may think of Wh as a subset of either S^3 or R^3. First let us check that $S^3 - Wh$ is contractible, for which we need $\pi_1 = 0$ and $H_2(S^3 - Wh; Z) = 0$. The fundamental group is generated by circles linking T_0. Consider $C_0 = (\text{point } \times \partial B^2) \subset S^1 \times B^2 = T_0$. Then C_0 contracts to a point in $S^3 - T_1$ and therefore in $S^3 - Wh$; this follows from Figure 4.2

Figure 4.2

since C_0 is obviously homotopic (although not isotopic) to a point in the complement of T_1. Similarly the linking circle C_k is homotopically trivial in the complement of T_{k+1}. But notice that these homotopies of C_i to a point cannot be made to miss $T_0^* = S^3 - T_0$, for the isotopy reversing the components in the symmetric link must affect T_0^*. This is important because it shows that $S^3 - Wh$ is not homeomorphic to R^3 because in R^3, a loop near ∞ can be contracted near ∞, whereas a loop C_k (which is near ∞ in $S^3 - Wh$ for large k) must hit T_0^* when contracted. This is made precise by the

DEFINITION: A space X is simply connected at ∞ if given any compact set K, there exists a compact set L with $K \subset L$, such that loops outside L can be shrunk outside K, i.e., $\pi_1(X - L) \to \pi_1(X - K)$ is the zero homomorphism.

$S^3 - Wh$ is not simply connected at ∞, for there is no L if K is chosen to be $T_0^* = S^3 - T_0$. Stallings' theorem [**Stall**] states that a contractible, simply connected at ∞, smooth n-manifold is diffeomorphic to R^n if $n \geq 5$. This is still unknown (it is equivalent to the Poincaré conjecture) in dimension 3 fifty years after Whitehead's mistake.

S^3/Wh, the quotient space when Wh is crushed to a point, is not a manifold at the quotient point (if it was, then clearly $S^3 - Wh$ would be simply connected at ∞). We will prove later that $(R^3/Wh) \times R$ is R^4 (XIII, §2).

The observant reader will have noticed that the torus T_k in Figure 4.1 is similar to the boundary of the slice disk used in defining the tower T_k in Figures 3.5, 3.6 and 3.7. In fact, if we identify T_0 with $B^2 \times B^1$ in $\partial(B^2 \times B^2)$, then each solid torus T_k is exactly a thickening of the dotted circle whose slice disk is subtracted from $B^2 \times B^2$ to get the kth-stage tower T_k. Moreover, the slice disk can be thought of as a thickened cone on the torus T_k. (Recall that these thickened slice disks are nested because a thickened slice for $T_0 = B^2 \times S^1$ is just $B^2 \times B^2$ in which we can imbed a slice for the solid torus T_1, and so on, by iteration.) Thus the k-stage tower T_k is diffeomorphic to $B^2 \times B^2$ minus a slice disk for the dotted circle equal to the solid torus T_k, which is just $B^2 \times B^2 -$ cone T_k. In the limit then, the simplest Casson handle CH is diffeomorphic to $(B^2 \times \text{ int } B^2) - (\text{cone } Wh)$. We have now shown Property 2 of Casson's Theorem in the simplest case.

In general, a Casson handle may be built with kinky handles with many kinks. This complicates matters only in that our construction of the Whitehead continuum must be generalized to allow for parallel copies of T_1 (called *replication*) (see Figure 4.3)

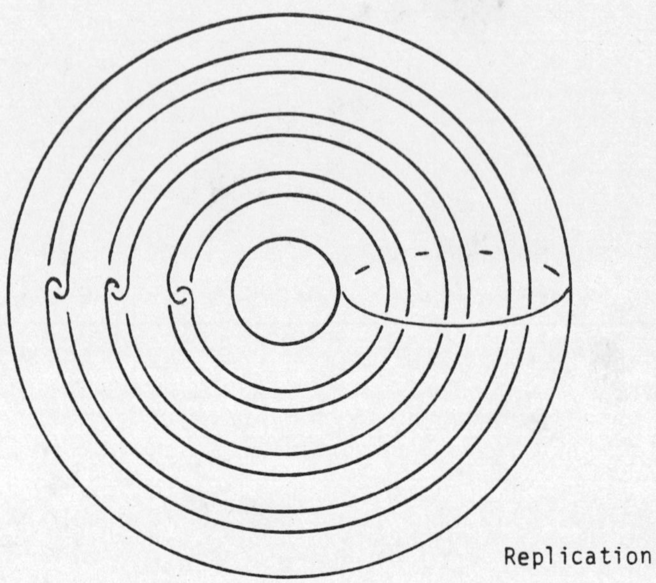

Figure 4.3

which corresponds to a kinky handle with many kinks (see Figure 4.4). For simplification, we will usually assume our kinky handles have only one kink.

This essentially finishes our description of Casson's work and lays the foundation for Freedman's work. Because the Casson handles are proper homotopy equivalent to standard handles there were no known invariants to distinguish them. It was known that they were standard topologically (smoothly) if one of the links in Figure 4.5 was topologically (smoothly) slice. It was known that the first three were not even topologically slice and the rest were doubtful. Now we know that they are topologically slice from the fifth one on, but that some family (allowing replication for extra kinks) is not smoothly slice.

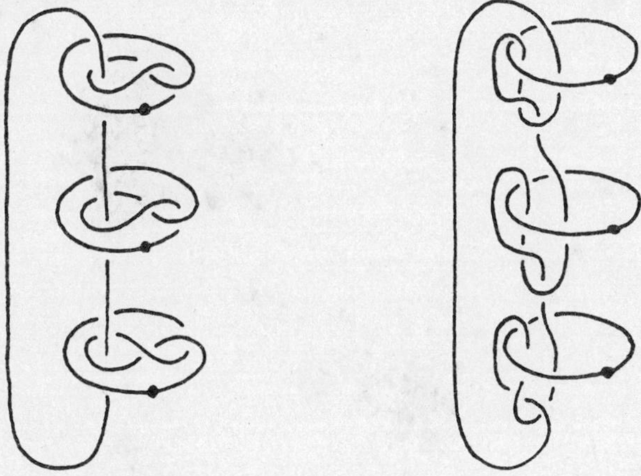

Figure 4.4

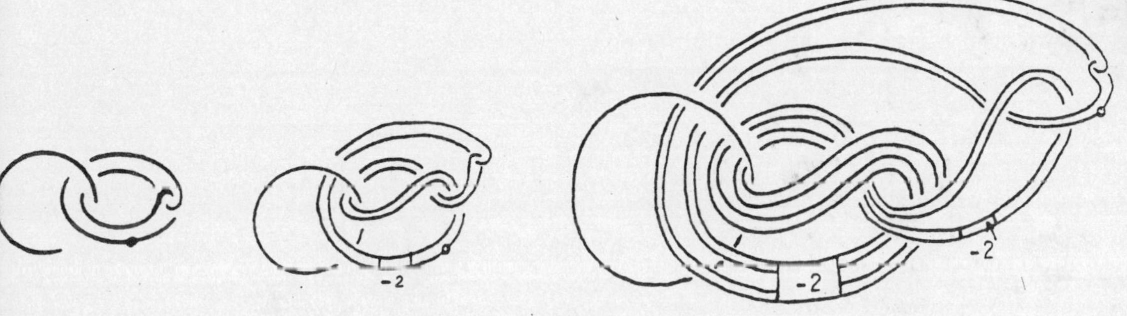

Further links in this sequence are obtained by taking the untwisted (hence the -2 twist to undo the other twists) double of the right component, and by allowing replications.

Figure 4.5

XIII. FREEDMAN'S WORK

§1. Decomposition.

We now have two relatively simple descriptions of the simplest Casson handles CH, i.e., Figure XII.3.7 and $B^2 \times B^2 -$ cone Wh. The latter suggests a clue which leads toward Freedman's proof that CH is pairwise homeomorphic to $(B^2 \times B^2, \partial_-)$, namely the fact (Shapiro–Bing) that $((B^2 \times S^1)/Wh) \times R$ is homeomorphic to $B^2 \times S^1 \times R$. Suppose that we could imbed some Casson handle CH_1 in our given one CH_0 (with $\partial_- CH_1 = \partial_- CH_0$) so that the frontier of CH_1, $(B^2 \times S^1) - Wh$, had closure in CH_0 equal to the one point compactification of $(B^2 \times S^1) - Wh$ which is just $(B^2 \times S^1)/Wh$. (This is analogous to imbedding $B^3 - \partial B^3$ in R^4 as the unit 3-sphere minus the north pole, and then observing that its closure is $B^3/\partial B^3 = S^3$.) This extra control on the imbedding of CH_1 in CH_0 is achieved by one of Freedman's technically hardest theorems (§3). But this only gives one copy of $(B^2 \times S^1)/Wh$ whereas we seem to need R copies, $((B^2 \times S^1)/Wh) \times R$, to conclude that CH_0 contains $(B^2 \times S^1/Wh) \times R = B^2 \times S^1 \times R$ which contains a 2-handle $B^2 \times B^1 \times B^1$ with the same attaching set ∂_-. Freedman cannot find R copies, but he does find a Cantor-set of copies of $(B^2 \times S^1)/Wh$ in CH_0, and then he strengthens the Shapiro–Bing theorem sufficiently to conclude that CH_0 is topologically an honest 2-handle.

There are two ingredients in Freedman's proof: one is classical differential topology, specifically, a sort of 4-dimensional handlebody theory (like the handlebody definition of CH) which leads to the Big Reimbedding Theorem (§3); the other is classical decomposition space theory as pioneered by R. H. Bing (§§1–2). These ingredients are joined by "the Design" which organizes Freedman's proof (§4).

A decomposition $\mathcal{D}$ of a topological space X is a partition of X into closed subsets, some of which are points and some of which are non-trivial. Our example is $X = S^1 \times B^2 \times R$ with closed subsets $\{Wh_t\} = \{Wh \times t\}$ for $t \in R$ and $\{p \times t\}$ for $t \in R$ and $p \in S^1 \times B^2 - Wh$ (the latter are trivial elements of $\mathcal{D}$). The quotient space of X where the closed sets of X are crushed to points is denoted $X/\mathcal{D}$. The principal question is when is the quotient map $X \xrightarrow{q} X/\mathcal{D}$ approximable by homeomorphisms (ABH); of course, this would imply that $X/\mathcal{D}$ is homeomorphic to X.

One's intuition can be improved by considering these examples and propositions:

1) Let $X = R^2$ and $\mathcal{D} = \{(x,y) \mid 0 \le x \le 1 \text{ and } y = 0\}$ and trivial elements; then $R^2 \xrightarrow{q} R^2/\mathcal{D}$, which simply crushes an interval in R^2 to a point, is ABH.

2) If X is an n-manifold and $\mathcal{D}$ has only one non-trivial element D_0, then D_0 is the intersection of a nested sequence of topological n-balls iff $X \to X/\mathcal{D}$ is ABH (the if part is easier, but both should be understood).

3) If $X \to X/\mathcal{D}$ is ABH, then no element of $\mathcal{D}$ can be a circle. (In fact, an element of $\mathcal{D}$ must be cell-like, which means that this element D is (i) compact and metric, and

(ii) imbeds in the Hilbert cube I^∞ such that D is null homotopic in any neighborhood U of D in I^∞.)

The main tool for showing that a decomposition is ABH is the Bing Shrinking Criterion (one says that $\mathcal{D}$ is shrinkable),

THEOREM (Bing) [**Edwards1**]. *For X a compact metric space, $q : X \to X/\mathcal{D}$ is $ABH \iff$ the following Bing Shrinking Criterion holds; given $\varepsilon > 0$, there exists a homeomorphism $h : X \to X$ such that*

1) $\sup_{x \in X} \operatorname{dist}(qh(x), q(x)) < \varepsilon$

2) $\operatorname{diameter}(h(D)) < \varepsilon$ *for all* $D \in \mathcal{D}$.

Edwards' proof of the important implication, $\Longleftarrow$, is so short and slick that it is worth quoting [**Edwards1**]. In the Baire space $C(X, X/\mathcal{D})$ of continuous maps with uniform topology, let $\mathcal{E}$ be the closure of the set $\{qh \mid h : X \to X \text{ is a homeomorphism}\}$. The Bing Shrinking Criterion amounts to saying that for any $\varepsilon > 0$, the open set of ε-maps in $\mathcal{E}$ (maps having all point inverses of diameter $< \varepsilon$), denoted $\mathcal{E}_\varepsilon$, is dense in $\mathcal{E}$. Hence $\mathcal{E}_0 \equiv \bigcap_{\varepsilon > 0} \mathcal{E}_\varepsilon$ is dense in $\mathcal{E}$ since $\mathcal{E}$ is a Baire space. Since $\mathcal{E}_0$ consists of homeomorphisms, $q \in \mathcal{E}$ is ABH. □

We need to apply the Bing Shrinking Criterion to $((B^2 \times S^1)/Wh) \times R$ which is not compact; there are ways around this difficulty, e.g., use a non-compact version of the criterion, or cross with S^1 to get a compact situation and later take the ∞-cyclic cover.

§2. $(R^3/Wh) \times R$.

THEOREM 1 [**A-R**]. *$(R^3/Wh) \times R$ is homeomorphic to R^4.*

PROOF: If we are crushing only one copy (not R) of the Whitehead continuum to a point, then we would shrink it by using the extra dimension to undo the clasp in T_{k+1} and isotope it to be small in T_k. If we do this for large enough k, then we shrink Wh small enough to satisfy 2) of the Criterion. Furthermore, each isotopy of T_{k+1} takes place in T_k, and since T_k is taken by the quotient map to a small (depending on k) neighborhood of the point $\{Wh\}$, it follows that qh can be arbitrarily close to q. (The reader may enjoy visualizing an isotopy $h_t : R^4 \to R^4$, $t \in [0, \infty)$ such that $h_\infty = \lim_{t \to \infty} h_t = q$. Let h_t, $t \in [k, k+1]$, be obtained by unclasping $h_k(T_{k+1})$ in $h_t(T \times [-1/2^k, 1/2^k])$ and then shrinking $h_t(T_{k+1})$ smaller than 2^{-k} in $h_k(T_k)$. Then $qh_t^{-1} : R^4 \to R^4/Wh$ has a limit qh_∞^{-1} which is a homeomorphism, and $(qh_\infty^{-1})h_t$ are homeomorphisms approximating q.)

The problem is to shrink R copies of Wh simultaneously. Here is the construction. In Figure 2.1, let $\mu : R^3 \to R$ be a continuous function which is zero outside T_0, which measures the angle between a point of T_1 and the line L (as indicated in Figure 2.1), and is a continuous extension on $T_0 - T_1$.

Let $\tilde{\mu} : R^4 \to R^4$ be defined by $\tilde{\mu}(x,t) = (x, t + \mu(x))$ for $x \in R^3$, $t \in R$. Then for any t, $\tilde{\mu}(T_1 \times t)$ is unclasped and may be shrunk to a "vertical" torus by a "rotation" of the form $\tilde{\rho}(x,t) = (\rho_t(x), t)$ where ρ_t rotates R^3 around the axis orthogonal to the page by an angle equal to $-t$ radius.

To shrink each $Wh \times t$ smaller than ε, we must use $\delta\tilde{\mu}$ where $\delta << \varepsilon$, and use the pair (T_k, T_{k+1}) instead of (T_0, T_1) where k is large enough so that $\tilde{\rho}(\delta\tilde{\mu})(T_{k+1} \times t)$ has a

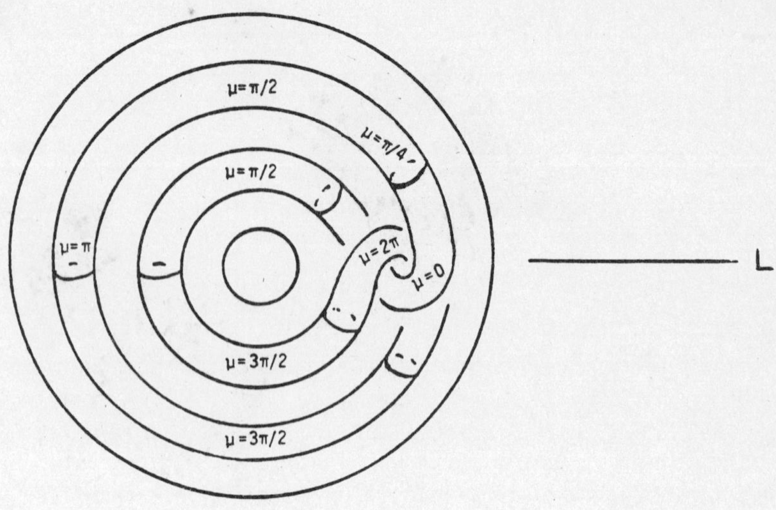

Figure 2.1

diameter less than ε (by the choice of δ) and $\tilde{\rho}(\delta\tilde{\mu})q$ is close enough to q (by the choice of k).

To "see" what happens, consider the thickened solid tori $T(n) = T_k \times [2\pi n-\gamma, 2\pi n+\gamma]$ where $n \in Z$ and γ is very small. Then $\tilde{\mu}(\bigcup_{n\in Z} T(n))$ is a chain since $\tilde{\mu}(T(n))$ links $\tilde{\mu}(T(n+1))$, which is coiled rather like a cord on a telephone (see Figure 2.2). The rotation $\tilde{\rho}$ simply unwinds or uncoils the chain so that it is vertical. The links $\tilde{\rho}(\delta\tilde{\mu})(T(n))$ have horizontal size depending on k and vertical length depending on δ.

The homeomorphism $(R^3/Wh) \times R \to R^4$ (call it Φ) should be contemplated because it (actually a far more complicated version) is the only non-differentiable part of Freedman's proof. Note that $\Phi(Wh \times R)$ is a wild arc in R^4, meaning that $\Phi(Wh \times R)$ does not have a neighborhood homeomorphic to $R \times B^3$. This is true because $R^4 - (Wh \times R)$ is not simply connected at ∞. In fact, $\Phi(Wh \times R)$ must intersect every horizontal 3-plane, $R^3 \times t$, in a Cantor set.

§3. Reimbedding Theorems.

Keeping in mind decomposition space techniques, we want to understand a given Casson handle CH_0 well enough to set up some kind of decomposition of it. Freedman "explores" CH_0 by imbedding a Cantor set of Casson handles inside it. For this we need reimbedding theorems.

Recall that an n-stage tower T_n (XII, §3) is a kinky handle (T_1) union kinky handles to kill $\pi_1(T_1)$ (this is T_2) union kinky handles to kill $\pi_1(T_2)$ (this gives T_3) and so on n times. Note that $\pi_1(T_n)$ is generated by loops, one for each double point, in the last stage of kinky handles; in fact, T_n is diffeomorphic to a boundary connected sum of $S^1 \times B^3$'s (just cancel handles from the left in a figure like Figure XII.3.7). Given an n-stage tower T_n^0 (e.g., the first n stages of CH_0), we achieve control by constructing

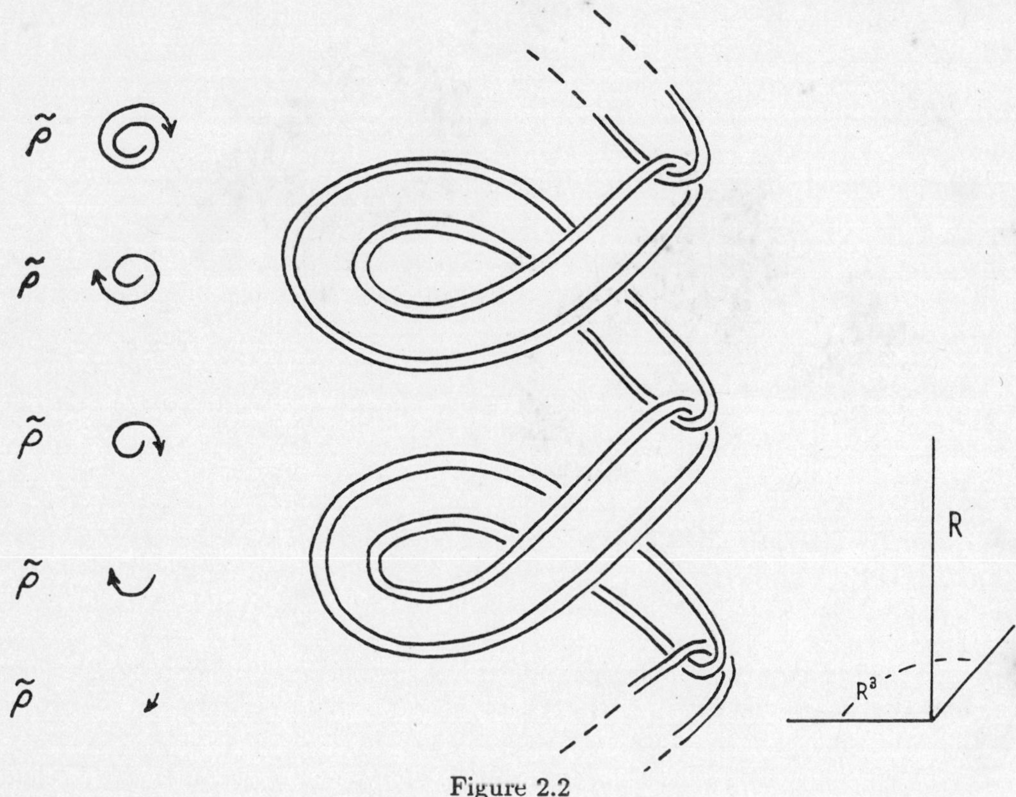

Figure 2.2

an imbedding of another n-stage tower T_n^1 in T_n^0 so that $\pi_1(T_n^1) \to \pi_1(T_n^0)$ is the zero homomorphism; even better, we arrange that the non-trivial loops in T_n^1 can be capped off by n-stage towers imbedding in T_n^0 (this is called mitosis). The first step is the

LITTLE REIMBEDDING THEOREM. *Every 3-stage tower T_3^0 contains another 3-stage tower T_3^1 (with same attaching circle) satisfying*

(a) *The first two stages agree, i.e., $T_2^1 = T_2^0$ and*
(b) $\pi_1(T_3^0 - T_3^1) \to \pi_1(T_3^0 - T_1^1)$ *is an isomorphism.*

We are being notationally sloppy, for T_2^1 is really a "thin" version of T_2^0. Then (b) means that if a loop in the complement of T_3^1 dies in the complement of T_1^1 then it must have been homotopically trivial already. This theorem is not hard to prove using finger moves (XII, §2) and is the first step in the

BIG REIMBEDDING THEOREM. *Every 4-stage tower T_4^0 contains another 4-stage tower T_4^1 such that*

(a) $T_2^1 = T_2^0$
(b) $\pi_1(T_4^0 - T_4^1) \to \pi_1(T_4^0 - T_1^1)$ *is an isomorphism*
(c) $\pi_1(T_4^1) \to \pi_1(T_4^0)$ *is zero.*

(Freedman used five stages here, but Gompf showed that five stages were unnecessary [G-S].)

The proof is long and technical, with many steps requiring finger moves; these create more non-trivial loops in T_4^1 which have to be killed, so it is not always clear that progress is being made. This theorem was proved by Freedman in 1978 and was the principal step in his construction of an exotic smooth structure on $S^3 \times R$ [**Freedman3**], [**Sieb1**].

A simple bootstrap operation strengthens the Big Reimbedding Theorem to:

MITOSIS. *Every five-stage tower T_5^0 contains a five-stage tower T_5^1 such that the loops generating $\pi_1(T_5^1)$ can be capped off by six-stage towers imbedded in T_5^0. (We require six-stage towers so that the first five stages of the caps are themselves homotopically trivial.)*

The five-stage tower should be thought of as the genes of a Casson handle, for inside it is a five-stage tower which can be capped off by a five-stage tower, and so on; thus a five-stage tower can replicate.

The proof is easy: by the Big Reimbedding Theorem, applied to the last four stages of T_5^0, there is another five-stage tower T_5^1 in T_5^0 (with $T_3^1 = T_3^0$). The loops generating $\pi_1(T_5^1)$ are homotopically trivial in the complement of T_5^1 (use (b)) and a calculation that a linking circle to T_2^1 dies in $T_5^0 - T_5^1$ (see Lemma 4.1 in [**Freedman1**]); by transversality the loops are capped off by immersed two-balls, i.e., by kinky handles, so we have T_6^1 in T_5^0.

Forget the first stage and apply this argument to get $T_{2-7}^2 \subset T_{2-6}^1$, or $T_7^2 \subset T_5^0$. Forget the first two stages and get $T_{3-8}^3 \subset T_{3-7}^2$, or $T_8^3 \subset T_5^0$. Eventually we get a $T_{11}^6 \subset T_5^0$ which is the same as a five-stage tower whose loops are killed by six-stage towers.

It is important to note here that mitosis gives us a way of imbedding a Casson handle so that the closure of its frontier is just $(B^2 \times S^1)/Wh$ (in the case of the simplest Casson handle; more generally we can have up to a Cantor set of Whitehead continua in $B^2 \times S^1$, each of which is crushed to a separate point, and here the closure adds a Cantor set to the frontier). To arrange this we require that the diameter of successive stages of CH tend to zero. So let T_{11}^1 be an 11-stage tower in any space, say some T_5^0, and observe that T_{6-10}^1 is homotopic to a collection of circles which can be homotoped, hence isotoped, inside ε-balls while fixing $\partial_-(T_5^1)$. Then inside T_{6-10}^1 find a T_{6-16}^2 and shrink T_{11-15}^2 while fixing $\partial_-(T_{6-10}^1)$. This ensures that successive stages have diameters going to zero while earlier stages are not stretched large. This technique allows us to assume that Casson handles have the right sort of frontiers whenever we wish, so we ignore the question from now on.

Freedman had essentially reached this point in 1978 [**Freedman3**]. The reimbedding theorems led to, and were motivated by, the existence of a smooth 4-manifold which is proper homotopy equivalent to but not diffeomorphic to $S^3 \times R$ (now we know it is homeomorphic to $S^3 \times R$). This was the first example of a "fake" smooth simply connected 4-manifold, although Cappell and Shaneson [**C-S**] in 1975 had found a fake RP^4 (homotopy equivalent but not diffeomorphic to RP^4). Casson had already shown in 1974 that there was either a fake $S^3 \times R$ or a fake end (in this case a smooth manifold proper homotopy equivalent to $S^2 \times S^2$-point with no smoothly imbedded 3-sphere near the missing point) but he was unable to determine which possibility held; now we know both exist.

We finish this section with a sketch of the exotic $S^3 \times R$ from [**Freedman3**] although it is not necessary for understanding the rest of this outline. Consider the Poincaré homology 3-sphere P^3; it bounds a simply connected smooth 4-manifold X^4 with intersection form E_8 [**K-Sch**] (any other homology 3-sphere bounding such a spin 4-manifold of index $\equiv 8(16)$ would do). $\pi_1(P^3)$, which is the binary icosahedral group, is normally generated by one element. If we surger that element in $P^3 \times [0,1]$, we obtain a smooth simply connected 4-manifold Y with $H_2(Y,Z) = Z \oplus Z$ because Y looks homologically like $(S^3 \times I) \sharp (S^2 \times S^2)$ (to see this, surger (i.e., remove $S^1 \times B^3$ and glue in $B^2 \times S^2$) a trivial circle in the interior of $S^3 \times I$).

We can represent the $Z \oplus Z$ in Y by a pair of Casson handles attached to in the boundary of a 4-ball in Y. For each Casson handle we use the Big Reimbedding Theorem to imbed a sequence of 5-stage towers $T_5^0 \subset T_5^1 \subset T_5^2 \subset \ldots$ such that $\pi_1(T_5^{k_1}) \to \pi_1(T_5^k)$ is zero and T_5^0 is the first five stages of the Casson handle. Then we remove the 4-ball union the intersection $\bigcap_{k=0}^{\infty} T_5^k$ from each Casson handle. This is proper homotopically like removing the wedge of two 2-spheres, and leaves a simply connected smooth 4-manifold $\tilde{Y}$ with boundary $P \cup -P$, the homology of $S^3 \times I$, and one end which is proper homotopy equivalent to the end of R^4.

Glue together a countable collection of $\tilde{Y}$'s along their boundaries P and $-P$ so that we get a manifold which is proper homotopy equivalent to $S^3 \times R$ with a countable number of points (e.g., $p \times n$, $p \in S^3$, $n \in Z \subset R$) removed. Finally, pick a transverse copy of P and join all the missing points on the right by an arc and remove it; similarly on the left. What remains is proper homotopy equivalent to $S^3 \times R$ but is not diffeomorphic to $S^3 \times R$ because there is a spanning Poincaré homology sphere. (If a diffeomorphism existed, we could compactify smoothly with two points and get S^4 with P smoothly imbedded; then P would bound a smooth contractible 4-manifold which could be added to X^4 to get a closed, smooth, almost parallelizable 4-manifold, contradicting Rochlin's theorem.)

§4. The Design.

Freedman's aim was to understand Casson handles and that evolved into an exploration of a Casson handle by an uncountable collection (the design) of imbedded Casson handles which have good frontiers in the sense we have previously explained.

Let CH_* be a given Casson handle, e.g., the simplest one. We imbed a 5-stage tower T_5^0 in CH_* and another 5-stage tower T_5^2 inside T_5^0 (the choice of superscripts will soon be clear; also all towers in this section will be 5-stage so we drop the 5). We kill the non-trivial loops in T^0 by adding a tower T^{00} in CH_* and we imbed another copy T^{02} inside T^{00} which also kills the loops of T^0. Similarly we kill the loops of T^2 by adding a tower T^{20} inside T^0, and we put another copy T^{22} inside T^0 (see Figure 4.1).

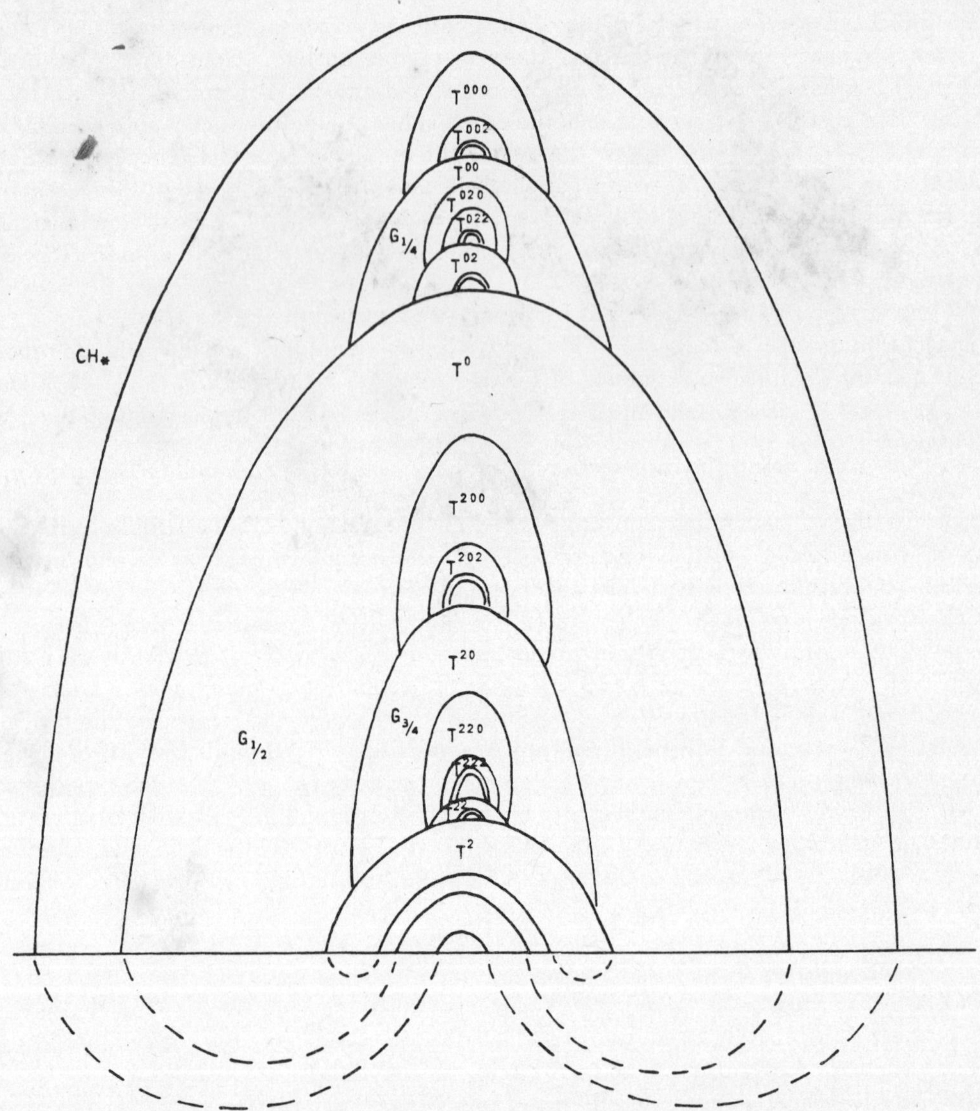

Figure 4.1 *The towers are drawn as* 1*-handles* ⌒ *but it should be remembered that they are more like* 2*-handles* $B^2 \times B^2$ *with connected boundary* $B^2 \times S^1$*, as indicated by dotted lines.*

We iterate this process, obtaining $T^{222} \subset T^{220}$ which lie in T^{20} and kill the loops of T^{22}; similarly with $T^{202} \subset T^{200}$ and $T^{022} \subset T^{020}$ and $T^{002} \subset T^{000}$. The pattern should be clear.

This gives a Casson handle for each element of the Cantor set C represented by base three decimals without any ones, i.e., $.n_1n_2n_3n_4\ldots$ where $n_i \in \{0,2\}$, for example, the Casson handle corresponding to .02020202... is obtained as the union of the 5-stage towers $T^0 \cup T^{02} \cup T^{020} \cup T^{0202} \cup T^{02020} \cup \ldots$. We are interested in the frontiers of these Casson handles (the boundary minus the attaching set ∂_-) which will be denoted F_γ, $\gamma \in C$. Recall once again that the closure of F_γ in CH_* is $(B^2 \times S^1)/Wh_\gamma$ where Wh_γ is a generalized Whitehead continuum which allows for replication (as in Figure XII.4.3) and where each component of Wh_γ is crushed to a separate point. If γ and γ' are close in $\mathcal{C}$, then F_γ and $F_{\gamma'}$ are pointwise close as sets since γ and γ' agree on their first N digits (N large). Thus the F vary continuously with respect to $\gamma \in \mathcal{C}$. (Note that as defined, $F_\gamma \cap F_{\gamma'} \neq \emptyset$, but we can make all the F_γ disjoint by using collar neighborhoods of ∂T_5 in T_5 to push F_γ slightly away from $F_{\gamma'}$.)

The F_γ, $\gamma \in \mathcal{C}$, "explore" CH_*, but they miss big gaps. For example, there is a gap between F_γ and $F_{\gamma'}$, $\gamma = .022222\ldots$ and $\gamma' = .200000\ldots$ which is not explored by any $F_{\gamma''}$; in fact, this gap corresponds to the first middle third which is deleted by constructing $\mathcal{C}$ and $\gamma \cup \gamma' = \partial[.1n_2n_3n_4\ldots] = \partial$[all numbers beginning .1]. Thus CH^* is the union of all F_γ, $\gamma \in \mathcal{C}$, and a countable number of gaps, G_β, where β runs over the middle thirds; these middle thirds correspond to the rational binary numbers where we write a binary number using $\{0,2\}$ and then $.02222\cdots = .20000\ldots$ is the rational $1/2$. So $\beta \in Q =$ binary rationals.

These gaps are not contractible, but can be shown to be the shape of a circle. It might be thought that we have explored very little—that the gaps are almost everything. Perhaps Freedman's boldest conception is that the gaps can be ignored, can be crushed to points. Well, not quite, for one cannot crush circles and get a manifold; so we crush a gap and a bit more (something like a 2-disk killing the circle) so that a contractible object is crushed.

To organize this incipient decomposition, we should reflect on the diagram in Figure 4.2.

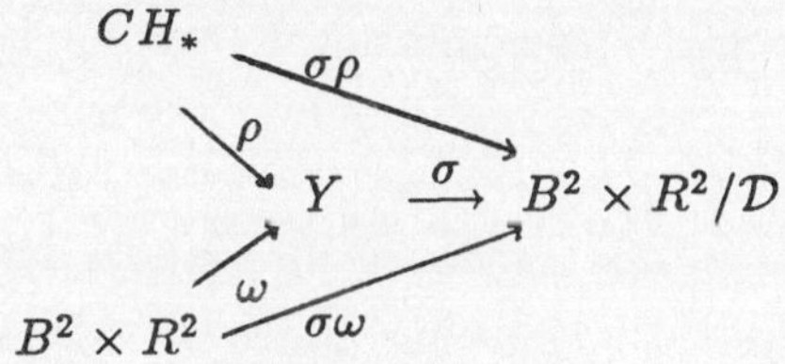

Figure 4.2

Let $B^2 \times R^2 = (B^2 \times S^1 \times R) \cup (B^2 \times 0)$ and crush each $Wh_t \times t$ to a point to get Y, where $Wh_t \in B^2 \times S^1$ and $t \in R$, and $Wh_t = Wh_\gamma$ if $t = \gamma \in C$ and Wh_t is chosen continuously with respect to t otherwise. Let ω be the quotient map.

Define a relation $\rho : CH_* \to Y$ which is a homeomorphism on the F_γ, $\gamma \in \mathcal{C}$, and a relation on the gaps G_β, $\beta \in Q$. More precisely, $\rho \mid F_\gamma$ is just the homeomorphism between F_γ and $((B^2 \times S^1)/Wh_\gamma) \times \gamma$; for $\rho \mid G_\beta$ let λ_β be the middle third of $[0,1]$ corresponding to β, let $\tilde{T}_k$ be some replicates of the k-th solid torus (in the definition of Wh) where $\beta = k/2^n$, and let ρ take G_β to the set $\tilde{T}_k \times \lambda_\beta$ in $(B^2 \times S^1) \times R$ in Y.

We cannot define σ to be the map which crushes $\tilde{T}_k \times \lambda_\beta$ to a point (points if $\tilde{T}_k$ has more than one component) for each $\beta \in Q$, because we would be crushing non-contractible sets. Since $\tilde{T}_k$ is a solid torus (or tori), we can find a way in Y (or even $B^2 \times R^2$) to imbed disjoint 2-balls, indexed by β, which cap off $\tilde{T}_k$ and then we can crush these larger, contractible sets to points. That roughly defines σ, and also the decomposition $\mathcal{D}$ which refers to the non-trivial point inverses of $\sigma\omega$.

What are the virtues of these maps and relations? First of all, the decomposition $B^2 \times R^2 \xrightarrow{\sigma\omega} (B^2 \times R^2)/\mathcal{D}$ is a decomposition on a well known space, not a mysterious object like CH_*. Furthermore, it differs from the classical $(B^2 \times R^2)/Wh \times R$ by crushing a well specified countable collection of sets, the non-trivial point inverses of σ. It can be shown (the proof is Edwards') by a much more technical, yet still classical, argument that $\mathcal{D}$ is shrinkable so that $\sigma\omega$ is ABH (approximable by homeomorphisms). Thus $(B^2 \times R^2)/\mathcal{D}$ is $B^2 \times R^2$.

Second, σ has crushed sets which contain $\rho(G_\beta)$ for all $\beta \in Q$; thus $\sigma\rho$ is a continuous map with only countably many non-trivial point inverses from CH_* to $B^2 \times R^2$. The problem that G_β is not contractible has been cleverly swept under the rug for $\rho(G_\beta)$ lies in a contractible set which is crushed by σ. Now we recall that CH_* is a subset of $B^2 \times B^2$ and hence of S^4, and that $B^2 \times R^2 \subset S^4$, and we extend $\sigma\rho$ to a continuous map $f : S^4 \to S^4$.

All we know (or need to know) about f is that it has countably many non-trivial point inverses whose diameters go to zero and whose images are nowhere dense. Remarkably, this is enough to prove that f is ABH. An easy step then shows that $\sigma\rho$ is ABH (rel ∂_-) so that CH_* is indeed homeomorphic to the standard handle $B^2 \times R^2$.

This finishes an increasingly sketchy outline of Freedman's main theorem, that Casson handles are topological handles. Since we can now imbed handles or $S^2 \times B^2$'s whenever the intersection form predicts (Theorem XII.3.1), we can now obtain the topological h-cobordism theorem and classification theorems as in high dimensions; all this is nicely explained in the introduction to [**Freedman1**].

For the details of a more modern proof of Freedman's theorem, and for all you ever wanted to know about topological 4-manifolds, see the forthcoming book of Freedman and Quinn, [**F-Q**].

XIV. EXOTIC R^4'S

It was morally clear from Casson's 1973-74 work ([**G-M**]) that if some Casson handles were topological handles but not smooth handles, then there would have to be exotic smooth structures on R^4. (This follows because there would be a set in B^4 which was topologically cellular but not smoothly cellular, that is, there would be open sets which were homeomorphic to R^4 but weren't smoothly R^4. See Edwards' remark, page 234 [**G-M**].) So when Freedman heard in March 1982 of Donaldson's theorem that any non-trivial definite form, say $-(E_8 \oplus (1))$, was not represented by a smooth 4-manifold, he knew that there ought to be an exotic R^4. The proof (Theorem 3 below) he gave produced an exotic R^4 that lay smoothly inside S^4, but required a compact counterexample to the smooth h-cobordism theorem which was not known until [**Don3**]. The proof that others found, assembling the ingredients of Casson, Donaldson and Freedman, produced an exotic R^4 that does not smoothly imbed in S^4. The author's version of this is Theorem 1 below.

After one exotic R^4 was found, the search began for more. It was soon realized that the open balls of sufficiently large radius inside the exotic R^4 provided an uncountable collection of possibly different exotic R^4's, but it was not until 1985 that they were shown to be different [**Taubes**].

Meanwhile Gompf found three exotic R^4's in 1982 [**Gompf1**] and then a countable collection [**Gompf2**] using the construction of Freedman and Taylor [**F-T**] of a universal exotic R^4 which contained all other exotic R^4's as open subsets. Then in fall 1984 during tea at MSRI, Gompf found an easy construction of countably many exotic R^4's which is given in Theorem 2 below.

THEOREM 1. *There exists an exotic smooth structure on R^4.*

PROOF: Recall our motivating example $M^4 = CP^2 \overset{10}{\#} (-CP^2)$ which has intersection form $-(E_8 \oplus 1) \oplus \begin{matrix} \alpha \\ \beta \end{matrix}\begin{pmatrix} 0 & 1 \\ 1 & 0 \end{pmatrix}$ with generators α and β with $\alpha \cdot \alpha = \beta \cdot \beta = 0$ and $\alpha \cdot \beta = 1$. We have seen (Theorem XII.3.1) that these two classes can be represented by topologically imbedded 2-spheres. In fact more is true: there exists a smooth manifold W^4, smoothly imbedded in M^4, where $W^4 = (\text{0-handle}) \cup (CH_1) \cup (CH_2)$ and the two (perhaps different) Casson handles are attached to  and boundary is deleted so that, by Property 3 of Theorem XII.3.1, W is homeomorphic to $S^2 \times S^2$-point. W is also a smooth submanifold of $S^2 \times S^2$, namely

$$\text{int}\{(\text{0-handle}) \cup (B^2 \times B^2 - \text{ cone } Wh_1) \cup (B^2 \times B^2 - \text{ cone } Wh_2)\}$$

for some generalized Whitehead continua Wh_1 and Wh_2 in $S^1 \times B^2$.

The topological core $S^2 \vee S^2$ of W in M^4 carries the homology classes α and β in $H_2(M^4; Z)$. The complement of this $S^2 \vee S^2$ in $S^2 \times S^2$ is proper homotopy equivalent to

R^4 (use the fact that $S^2 \vee S^2$ has a neighborhood homeomorphic to $S^2 \times S^2$-point to see that the end is $S^3 \times R$), hence by Freedman's topological proper h-cobordism theorem [**Freedman1**] is homeomorphic to R^4. This R^4, being an open subset of $S^2 \times S^2$, inherits a smooth structure Σ which we show is exotic.

Suppose not. Then there are arbitrary large "round" smooth 3-spheres in R^4_Σ, large enough to surround any compact set in this R^4. In particular there would be a smooth 3-sphere large enough to contain the closed, compact set $K = S^2 \times S^2 - W$, so therefore this 3-sphere lies smoothly in W, but misses $S^2 \vee S^2$. Then this 3-sphere also lies smoothly in M^4 with α and β (represented by $S^2 \vee S^2$) on one side and $-(E_8 \oplus (1))$ on the other. Cut M^4 along this smooth 3-sphere and glue in a 4-ball to get a contradiction to Donaldson's Theorem (III, §3). Figure 1 may help visualize this construction, except that W^4 is far more convoluted than can be drawn. □

REMARK 1: The characterizing property of this R^4_Σ is the compact set K which cannot be surrounded by any smoothly imbedded 3-sphere. It also cannot be surrounded by any homology 3-sphere which bounds a smooth acyclic 4-manifold because then the same construction would contradict Donaldson's theorem.

REMARK 2: The same contradiction is achieved if an open neighborhood U of K, U homeomorphic to R^4, was smoothly imbedded in a smooth homotopy 4-sphere Q^4. Then $U - K$ is the end of W. We delete from M^4 the closed set $W^4 - (U - K)$. We can add on the smooth open manifold $Q - K$ along $U - K$. This gives a smooth manifold whose form is $-(E_8 \oplus 1)$, contradiction.

THEOREM 2 (Gompf). *R^4 has countably many exotic smooth structures.*

PROOF: We begin with the definition of the end-connected-sum of two copies of R^4: choose a smooth arc γ in R^4 from a point p to infinity. Any two such are diffeotopic. γ has a tubular neighborhood and its boundary is a smooth copy of R^3 properly imbedded in R^4, namely $\gamma \times S^2$ union $p \times B^3$ with corners rounded. Now take two copies of R^4 and form $R^4 \overset{\text{end}}{\#} R^4$ by choosing arcs γ_1 and γ_2 in each, throwing away the interiors of their tubular neighborhoods, and gluing the two copies of R^3 together by an orientation reversing diffeomorphism. It is not hard to see that the diffeomorphism type of $R^4 \overset{\text{end}}{\#} R^4$ is independent of the various choices.

From Theorem 1 we have an exotic structure Σ on R^4 characterized by a compact set K which cannot be surrounded by a smooth 3-sphere. We can imbed K inside a ball of radius k, kB^4. Outside kB^4 choose two disjoint smooth (in R^4_Σ) arcs γ_1 and γ_2 from p_1 to ∞ and p_2 to ∞. For each, choose tubular neighborhoods and then R^3's which are disjoint (and smooth in R^4_Σ). Now form the end-connected-sum, E, of a countable collection of R^4_Σ's, in which the ith copy of R^4_Σ is glued to the $(i-1)$-copy along the R^3 of γ_1 and is glued to the $(i+1)$-copy along the R^3 of γ_2.

It is easy to see that E is homeomorphic to R^4, for the arcs are ambiently isotopic to $\{(x_1, 0, 0, 0) \in R^4 \mid x_1 \leq -1\}$ and $\{(x_1, 0, 0, 0) \in R^4 \mid 1 \leq x_1\}$. It contains a countable number of copies of K, one for each R^4_Σ. It is not hard to choose topological coordinates on E (that is, a homeomorphism to R^4) so that the first copy of K lies in int B^4, the second in int $2B^4 - B^4$, the ith copy in int $iB^4 - (i-1)B^4$, and so on.

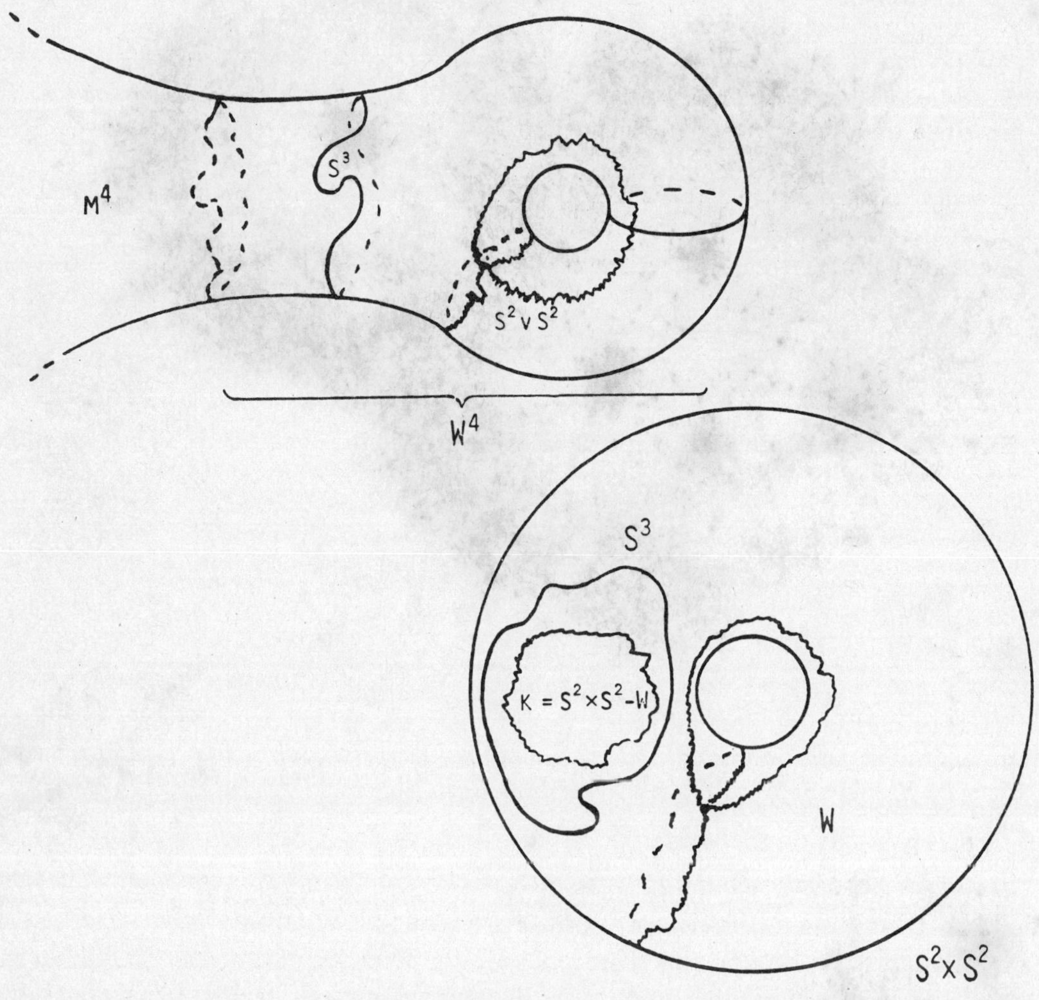

Figure 1

Now we show that the open balls $\text{int}(iB^n)$, $i = 1, 2, 3, \ldots$ with the smooth structure they inherit from E, form a countable, distinct collection of exotic R^4's. Suppose not, i.e. suppose for $i < j$ that there is a diffeomorphism $f : \text{int } iB^4 \to \text{int } jB^4$. Let A be the open annular region which f maps onto int $jB^4 - (j-1/2)B^4$. Then "furl" the smooth manifold (int $jB^4 - iB^4) \cup A$ by using f to glue the two ends together; the smooth result, N^4, is homotopy equivalent (in fact homeomorphic) to $S^3 \times S^1$. Furthermore, N contains $j - i > 0$ copies of K. It is not hard to find a smooth circle in N which misses at least one of the copies of K. Surgering this circle produces a manifold Q^4 which is homotopy equivalent to S^4 and contains a copy of K. But this contradicts Remark 2 after Theorem 1 because K, in fact kB^4, lies in a homotopy S^4. □

THEOREM 3 (Casson and Freedman). *There exists an exotic R^4_Θ which imbeds smoothly in S^4.*

In 1976 Casson (Lecture 3 in [**G-M**]) described a smooth 5-dimensional h-cobordism between compact 4-manifolds and showed that they "differed" by two proper homotopy R^4's (the construction is given below). Freedman knew, as an application of his proper h-cobordism theorem that the proper homotopy R^4's were R^4; after hearing of Donaldson's work in March 1983, Freedman realized there should be exotic R^4's and, to find one, he produced the second part of the construction below involving the smooth imbedding of the proper homotopy R^4's in S^4. Unfortunately, it was necessary to have a *compact* counterexample to the smooth h-cobordism conjecture, and Donaldson did not provide this until 1985 [**Donaldson3**]. This counterexample was a smooth h-cobordism Y^5 between the rational surface $CP^2 \sharp 9(-CP^2)$ and its logarithmic transform $L = L(2,3)$ known as the Dolgachev surface [**H-K-K**]. Y^5 is not a smooth product because $CP^2 \sharp 9(-CP^2)$ and $L(2,3)$ are not diffeomorphic.

(Incidentally, as soon as an exotic R^4, R^4_Σ, was found as in Theorem 1, there was an exotic, but non-compact, proper h-cobordism between R^4_Σ and R^4, obtained from $R^4_\Sigma \times I$ by removing from $R^4_\Sigma \times 1$ everything but a coordinate chart R^4.)

PROOF: Consider Y^5 as a handlebody built on L. As before, we can cancel the 0, 1, 4 and 5-handles. After handle slides, the remaining 2 and 3-handles occur in algebraically cancelling pairs. In higher dimensions the Whitney trick (Chapter XII, §1) can be used to make these pairs cancel geometrically so that no handles are left and Y is a product. In dimension 4 this can only be done topologically [**Freedman1**].

For simplicity of notation, we will use the fact (without proof) that Y can be constructed with just one 2-handle h_2 and one 3-handle h_3. Since h_2 is added to a simply connected 4-manifold L, it follows that the middle level of Y is $L \sharp S^2 \times S^2$; adding h_3 (upside down) to the rational surface gives $2CP^2 \sharp 10(-CP^2)$ which must also be $L \sharp S^2 \times S^2$. (Turning the argument around, we can see how to construct Y by seeing how to add one 2-handle to L so that it breaks apart into $2CP^2 \sharp 10(-CP^2)$. This can be done by adding the 2-handle so as to cancel the 1-handle in L (see page 75 [**H-K-K**]).)

After adding h_2 to L, the new boundary is the middle level of Y (modulo some collars), $Y_{1/2}$, and the cosphere of h_2 ($\partial(p \times B^3)$ in $B^2 \times B^3$) is a smooth 2-sphere S_2 in $Y_{1/2}$. The attaching 2-sphere S_3 of h_3 also lies in $Y_{1/2}$. S_2 and S_3 intersect algebraically once but geometrically $2k+1$ times, $k > 0$. For simplicity, we assume $k = 1$, so that $S_2 \cap S_3 = p_0 \cup p_1 \cup p_2$ (see Figure 2). Note that $S_2 \cdot S_2 = S_3 \cdot S_3 = 0$.

Figure 2

Associated with this handlebody structure on Y is a Morse function whose gradient vector field integrates to a smooth product structure on Y away from the set X of ascending and descending manifolds of the two critical points. In particular, $Y_{1/2} - (S_2 \cup S_3)$ flows up and down to give the smooth product structure on $Y - X$.

If we think of p_1 and p_2 as the unwanted pair of intersections (where the sign of p_i is $(-1)^i$), then, working in $Y_{1/2}$, we can imbed a toplogical Whitney disk which lies in a smoothly imbedded Casson handle, CH_1, whose boundary is a circle consisting of arcs joining p_1 to p_2 in S_2 and S_3. Furthermore we can smoothly imbed in $Y_{1/2}$ an auxiliary Casson handle, CH_2, whose attaching circle consists of two arcs joining p_0 to p_1 in S_2 and S_3 (see the schematic picture in Figure 3).

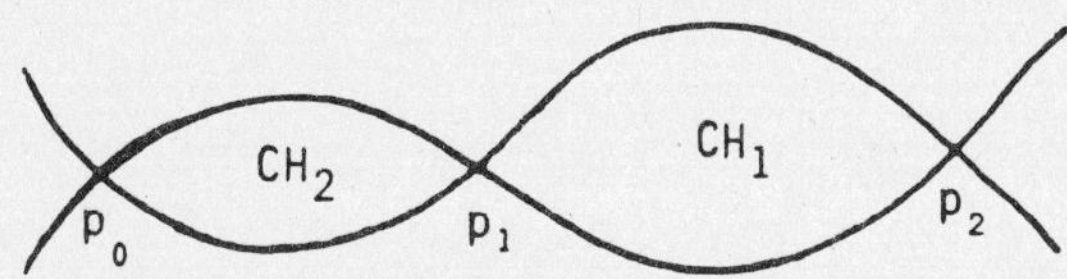

Figure 3

To construct CH_1 and CH_2, it is necessary (see Chapter XII, §2) to arrange that $\pi_1(Y_{1/2} - S_2 - S_3) = 0$ and to observe that there are dual classes, namely the characteristic tori at the p_i, to the proposed Whitney disks. To arrange that $\pi_1(Y_{1/2} - S_2 - S_3) = 0$, we use the fact that $\pi_1(Y_{1/2} - S_2) = \pi_1(L - \partial h_2) = 0$ so that conjugates of a meridian μ_3 to S_3 generate $\pi_1(Y_{1/2} - S_2 - S_3)$. Similarly, it is generated by conjugates of a meridian μ_2 to S_2. Since $H_1(Y_{1/2} - S_2 - S_3; Z) = 0$ it follows that $\pi_1(Y_{1/2} - S_2 - S_3)$ is equal to its commutator subgroup, and therefore is generated by elements of the form $[\mu_2^g, \mu_3^h] = [\mu_2, \mu_3^{g^{-1}h}]^{g^{-1}}$. Such elements can be killed by the right finger move (see Chapter XII, §2), that is we isotop the attaching map S_3 of h^3, increasing the number of pairs of points of intersection in $S_2 \cap S_3$ so as to make the complement simply connected. But from now on, we ignore these extra points.

In the middle level $Y_{1/2}$, let U be an open tubular neighborhood of $S_2 \cup S_3$. Adding the two Casson handles CH_1 and CH_2 to U give an open 4-manifold W^4 which is homeomorphic to an open tubular neighborhood of $S^2 \vee S^2$ in $S^2 \times S^2$ (since CH_1 and CH_2 are homeomorphic to $B^2 \times R^2$ and the attaching maps are smoothly standard, p_1 and p_2 and CH_1 and CH_2 can be amalgamated into a large open 4-ball around p_0).

Call Z the part of the h-cobordism Y which lies above and below W. Z is a smooth product outside the compact set X, hence near its end. Smoothly, Z has only the handles h_2 and h_3; these can be cancelled topologically because we can use CH_1 as a Whitney disk to eliminate the pair of double points p_1 and p_2. Thus Z is a topological product; in fact the smooth product structure on $Y - Z$ extends to a topological product over Z. Furthermore, Z must be homeomorphic to $R^4 \times I$ because Z is formed from the open tubular neighborhood W of $S^2 \vee S^2$ by adding 3-handles to each S^2 (one down, one up), as in Figure 4.

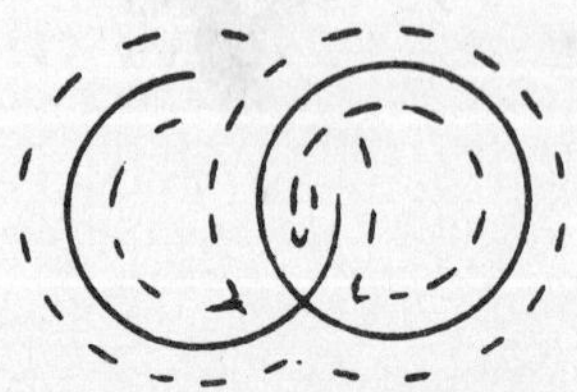

Figure 4

The components of ∂Z, Z_0 and Z_1, inherit smooth structures from $\partial Y = L \cup CP^2 \,\sharp\, 9(-CP^2)$; call these structures Θ_0 and Θ_1. Suppose Θ_0 is standard (there is a similar argument if Θ_1 is standard). Then there is a sufficiently large 4-ball of radius ρ, ρB^4, in $Z_0 = R^4_{\Theta_0}$ so that over $Z_0 - (\rho - 1)B^4$, Y is a smooth product (this follows if $(\rho - 1)B^4$ contains $X \cap L$). In fact, Y is a smooth product over $L -$ int ρB^4 with "vertical" boundary $\rho S^3 \times I$. If $\rho S^3 \times 1$ bounded a smooth 4-ball in $Z_1 \subset CP^2 \,\sharp\, 9(-CP^2)$, then, since all diffeomorphisms of S^3 extend over B^4, we would have constructed a diffeomorphism between L and $CP^2 \sharp 9(-CP^2)$. Thus either $\rho S^3 \times 1$ does not bound a smooth 4-ball in Z_1 or Θ_0 is exotic. This part of the argument is essentially in Lecture 3 in [G-M]; Freedman then shows that $\rho S^3 \times 1$ does bound a smooth 4-ball.

Next we show that Z smoothly imbeds in $S^4 \times I$, so of course Z_0 lies in $S^4 \times 0$ and Z_1 in $S^4 \times 1$. Consider the smooth product $S^4 \times I$ as an h-cobordism with one cancelling pair of 2 and 3-handles. Thus in the middle level we see a pair of 2-spheres (the cosphere S_2 of the 2-handle and attaching 2-sphere S_3 of the 3-handle) which meet transversely in one point p_0. In a neighborhood of p_0, isotop S_3 so that $S_2 \cap S_3 = p_0 \cup p_1 \cup p_2$ and there are two smooth Whitney disks D_1 and D_2 playing the roles that CH_1 and CH_2 did earlier. Inside smooth thickenings of D_1 and D_2, namely $D_1 \times R^2$ and $D_2 \times R^2$, we can smoothly imbed CH_1 and CH_2; in fact CH_i smoothly equals $D_i \times B^2$ minus a generalized cone on a Whitehead continuum. Thus the entire smooth manifold Z is smoothly imbedded in $S^4 \times I$.

Furthermore, as we saw for Y, there is an associated Morse function on $S^4 \times I$ whose gradient vector field integrates to a smooth product above and below $(S^4 \times 1/2) - (S_2 \cup S_3)$. Thus the smooth product structure on $Z - X$ coincides with the smooth product structure on $Z - X$ as a subset of $S^4 \times I$.

Recall that ρB^4 is smoothly imbedded in Z_0 and hence in $S^4 \times 0$. The complement of a smooth n-ball in S^n is always a smooth n-ball (because the smooth n-ball may be ambiently isotoped to a standard round n-ball), so $S^4 \times 0 - \rho B^4$ is a smooth 4-ball. Moving up the smooth product structure, we see that in $S^4 \times 1$, $\rho S^3 \times 1$ bounds a smooth 4-ball on the outside, hence in Z_1. Thus $R^4_{\Theta_0}$ is exotic. □

ADDENDUM: The smooth structures Θ_0 and Θ_1 on R^4 are equivalent, that is $R^4_{\Theta_0}$ is diffeomorphic to $R^4_{\Theta_1}$. Then L can be obtained by "surgery" on $R^4_{\Theta_1}$ from $CP^2 \sharp 9(-CP^2)$.

PROOF: We arrange a $Z/2$-symmetry on W as follows. S_2 and S_3 meet in three points and there is a diffeomorphism switching S_2 and S_3. When the Casson handles CH_1 and CH_2 are being constructed, there is always the freedom to add extra kinks (inside a small ball) at any stage, so if, say, CH_1 has k fewer kinks than CH_2 at the ith stage, then we add k extra kinks to CH_1. Doing this at each stage, we get CH_1 to be abstractly diffeomorphic to CH_2, although they may be imbedded very differently in $Y_{1/2}$. This diffeomorphism between CH_1 and CH_2 can be easily extended to an involution of W switching S_2 and S_3; after surgery on S_2 and S_3 by h_2 and h_3, the involution gives a diffeomorphism between the two boundary components of Z, $R^4_{\Theta_0}$ and $R^4_{\Theta_1}$, which we now refer to as R^4_{Θ}.

We have seen that $Y - X$ is a smooth product between $L - X_0$ and $CP^2 \sharp 9(-CP^2) - X_1$ and that $L = (L - X_0) \cup_{h_0} R^4_\Theta$ and $CP^2 \sharp 9(-CP^2) = (CP^2 \sharp 9(-CP^2) - X_1) \cup_{h_1} R^4_\Theta$ where h_0 and h_1 are diffeomorphisms on the overlaps. But then we can "surger" $CP^2 \sharp 9(-CP^2)$ by removing R^4_Θ and gluing it back on by the diffeomorphism h_0 to obtain L.

APPENDIX: THE ARF INVARIANT

Our discussion of the Arf invariant is lifted directly from [**R-S**, Appendix]. Let V be a vector space over $Z/2$ equipped with an inner product, i.e. $x \cdot y \in Z/2$, $x \cdot y = y \cdot x$, and given a linear map $\lambda : V \to Z/2$, there exists $y \in V$ such that $\lambda(x) = x \cdot y$ for all $x \in V$. $V = H_1(F^2; Z/2)$ with the standard intersection form is the obvious example.

Let $q : V \to Z/2$ be a quadratic function, i.e. q must satisfy

$$q(x+y) = q(x) + q(y) + x \cdot y \quad (2) \quad \text{for all } x, y \in V. \tag{*}$$

Note that applying (*) to $0+0$ and to $x+x$ shows that $q(0) = 0$ and $x \cdot x = 0$. Choose x and a dual y to x so that $x \cdot y = 1$ (y corresponds to the linear map which sends x to 1 and all other elements of V to zero). Then the hyperbolic pair $H = \begin{matrix} x \\ y \end{matrix}\begin{pmatrix} 0 & 1 \\ 1 & 0 \end{pmatrix}$ defines a unimodular subspace and hence an orthogonal complement $H^{\perp}$. We continue to split off hyperbolic pairs until nothing is left and we have shown that $V = H_1 \oplus H_2 \oplus \cdots \oplus H_n$.

It is easy to verify using (*) that on the three non-zero elements of H, x, y, and $x+y$, q is either always one (this case is called $H^{1,1}$) or q is zero on two of the elements and one on the third (this case is called $H^{0,0}$ because we can choose a basis of two elements on which q is zero). Thus $(V, q) = \overset{r}{\oplus} H^{0,0} \overset{s}{\oplus} H^{1,1}$, where $r + s = 2n$.

Given a basis x_1, y_1, x_2, y_2 for $H^{0,0} \oplus H^{0,0}$ we can choose a new basis $x_1 + y_1 + x_2$, $x_1 + y_1 + y_2$, $x_1 + x_2 + y_2$, $y_1 + x_2 + y_2$ on each element of which q is 1, so it follows that $H^{0,0} \oplus H^{0,0} \cong H^{1,1} \oplus H^{1,1}$. Thus (V, q) is isomorphic to either $\overset{n}{\oplus} H^{0,0}$ or $\overset{n-1}{\oplus} H^{0,0} \oplus H^{1,1}$ depending on whether s is even or odd.

Finally, $\overset{n}{\oplus} H^{0,0}$ is not isomorphic to $\overset{n-1}{\oplus} H^{0,0} \oplus H^{1,1}$ because out of the 2^{2n} elements in V, q is zero on $2^{2n-1} + 2^{n-1}$ of them in $\overset{n}{\oplus} H^{0,0}$ and is zero on $2^{2n-1} - 2^{n-1}$ of them in $\overset{n-1}{\oplus} H^{0,0} \oplus H^{1,1}$.

We define the Arf invariant of (V, q), $\mathrm{Arf}(V, q) \in Z/2$, to be zero if $(V, q) \cong \overset{n}{\oplus} H^{0,0}$ and one if $(V, q) \cong \overset{n-1}{\oplus} H^{0,0} \oplus H^{1,1}$. Note that the Arf invariant is additive under direct sum.

REFERENCES

[A-K1] Akbulut, S. and Kirby, R., *Branched covers of surfaces in 4-manifolds*, Math. Ann. **252** (1980), 111–131.

[A-K2] ________________, *Exotic involutions of S^4*, Topology **18** (1979), 75–81.

[A-K3] ________________, *A potential counterexample in dimension 4 to the Poincaré conjecture, the Schoenflies conjecture and the Andrews–Curtis conjecture*, Topology **24** (1985), 375–390.

[A-K4] ________________, *Mazur manifolds*, Mich. Math. J. **26** (1979), 259–284.

[A-R] Andrews, J. and Rubin, L., *Some spaces whose product with E is E^4*, Bull. A.M.S. **71** (1965), 675–677.

[Brown] Brown, M., *A proof of the generalized Schoenflies theorem*, Bull. A.M.S. **66** (1960), 74–76.

[C-H] Casson, A. M. and Harer, J., *Some homology lens spaces which bound rational homology balls*, Pac. J. Math. **96** (1981), 23–36.

[C-S] Cappell, S. and Shaneson, J., *Some new 4-manifolds*, Ann. Math. **104** (1976), 61–72.

[Cerf1] Cerf, J., *La stratification naturelle des espaces fonctions differentiables réeles et la thèoréme de la pseudoisotopie*, Publ. Math. I.H.E.S. **39** (1970).

[Cerf2] ______, *Sur les diffeomorphismes de la sphere de dimension trois ($\Gamma_4 = 0$)*, Lecture Notes in Math. **53** (Springer, Berlin, 1968).

[Don1] Donaldson, S., *An application of gauge theory to 4-dimensional topology*, J. Diff. Geom. **18** (1983), 279–315.

[Don2] ____________, *Connections, cohomology, and the intersection forms of 4-manifolds*, J. Diff. Geom. **24** (1986), 275–341.

[Don3] ____________, *Irrationality and the h-cobordism conjecture*, J. Diff. Geom. **26** (1987), 141–168.

[Edwards1] Edwards, R. D., *The topology of manifolds and cell-like maps*, Proc. International Congress of Mathematicians, Helsinki (1978), 111–127.

[Edwards2] ______________, *The solution of the 4-dimensional annulus conjecture (after Frank Quinn)*, in "4-Manifolds", Contemporary Math. **35** (1984), 211–264.

[F-R] Fenn, R. and Rourke, C., *On Kirby's calculus of links*, Topology **18** (1979), 1–15.

[F-S1] Fintushel, R. and Stern, R., *$SO(3)$ connections and the topology of 4-manifolds*, J. Diff. Geom. **20** (1984), 523–539.

[F-S2] ________________, *Pseudofree orbifolds*, Ann. Math. **122** (1985), 335–364.

[F-U] Freed, D. and Uhlenbeck, K., "Instantons on 4-Manifolds," MSRI Pub. 1, Springer-Verlag, New York, 1984.

[Freedman1] Freedman, M. H., *The topology of 4-dimensional manifolds*, J. Diff. Geom. **17** (1982), 357–453.

[Freedman2] ______________, *The disk theorem for 4-dimensional manifolds*, Proc. International Congress of Mathematicians, Warsaw (1983), 647–663.

[Freedman3] ______________, *A fake* $S^3 \times R$, Ann. Math. **110** (1979), 177–201.

[F-K] Freedman, M. and Kirby, R., *A geometric proof of Rohlin's theorem*, Proc. Symp. Pure Math. **32** (1978), 85–97.

[F-Q] Freedman, M. and Quinn, F., "Topology of 4-Manifolds."

[F-T] Freedman, M. H. and Taylor, L. R., *A universal smoothing of four-space*, J. Diff. Geom. **24** (1986), 69–78.

[F-M1] Friedman, R. and Morgan, J., *On the diffeomorphism types of certain algebraic surfaces, I, II*, J. Diff. Geom. (1988).

[F-M2] ______________, *Algebraic surfaces and 4-manifolds: some conjectures and speculations*, Bull. A.M.S. (1988).

[G-R] Gerstenhaber, M. and Rothaus, O. S., *The solution of sets of equations in groups*, P.N.A.S. USA **48** (1962), 1531–1533.

[Gluck] Gluck, H., *The embeddings of 2-spheres on the 4-sphere*, Bull. A.M.S. **67** (1967), 586–589.

[Gompf1] Gompf, R., *Three exotic* R^4*'s and other anomalies*, J. Diff. Geom. **18** (1983), 317–328.

[Gompf2] __________, *An infinite set of exotic* R^4*'s*, J. Diff. Geom. **21** (1985), 283–300.

[G-S] Gompf, R. and Singh, S., *On Freedman's reimbedding theorems*, in "4-Manifolds", Contemp. Math. **35** (1984), 277–309.

[G-M] Guillou, L. and Marin, A., *A la Recherche de la Topologie Perdue*, Progress in Math. **62** (Birkhäuser, Boston, 1986).

[Habegger] Habegger, N., *Une varieté de dimension 4 avec forme d'intersection paire et signature-8*, Comm. Math. Helv. **57** (1982), 22–24.

[H-P] Haefliger, A. and Poenaru, V., *La classification des immersions combinatoires*, Publ. Math. I.H.E.S. **23** (1964), 79–91.

[H-K-K] Harer, J., Kas, A. and Kirby, R., *Handlebody decompositions of complex surfaces*, Memoirs A.M.S. **62** (1986), p. 350.

[Herbert] Herbert, R., *Multiple points of immersed manifolds*, Memoirs A.M.S. **34** (1981), p. 250.

[Hirsch1] Hirsch, M., *Immersions of manifolds*, Trans. A.M.S. **93** (1959), 242–276.

[Hirsch2] __________, *The embedding of bounding manifolds in Euclidean space*, Ann. Math. **74** (1961), 494–497.

[Hirsch3] __________, "Differential Topology," Springer-Verlag, New York, 1976.

[Hirze] Hirzebruch, F., "New Topological Methods in Algebraic Geometry," Springer-Verlag, New York, 1978.

[H-N-K] Hirzebruch, F., Neumann, W. D. and Koh, S. S., "Differentiable Manifolds and Quadratic Forms," Marcell Dekker, New York, 1971.

[Hughes] Hughes, J., *Ph.D. thesis*, Berkeley (1982).

[Kaplan] Kaplan, S., *Constructing 4-manifolds with given almost framed boundaries*, Trans. A.M.S. **254** (1979), 237–263.

[K-M] Kervaire, M. and Milnor, J., *On 2-spheres in 4-manifolds*, Proc. Nat. Acad. Science USA **47** (1961), 1651–1657.

[**Kirby1**] Kirby, R., *A calculus for framed links in* S^3, Inv. Math. **45** (1978), 36–56.

[**Kirby2**] ________, *Problems in low dimensional manifold theory*, Proc. Sym. Pure Math. **32** (1978), 273–312.

[**Kirby3**] ________, *4-manifold problems*, in "4-Manifolds", Contemp. Math. **35** (1984), 513–528.

[**K-Sch**] Kirby, R. and Scharlemann, M., *Eight faces of the Poincaré homology 3-sphere*, "Geometric Topology", ed. J. C. Cantrell (Academic Press, New York, 1979), 113–146.

[**K-S**] Kirby, R. and Siebenmann, L., *Foundational essays on triangulations and smoothings of topological manifolds*, Ann. Math. Studies **88** (Princeton University Press, 1977).

[**K-T**] Kirby, R. and Taylor, L., *Spin and Pin structures on low dimensional manifolds.*

[**Kuga**] Kuga, K., *Representing homology classes of* $S^2 \times S^2$, Topology **23** (1984), 133–137.

[**L-P**] Laudenbach, F. and Poenaru, V., *A note on 4-dimensional handlebodies*, Bull. Math. Soc. France **100** (1972), 337–344.

[**Lickorish**] Lickorish, W. R. B., *A representation of orientable, combinatorial 3-manifolds*, Ann. Math. **76** (1962), 531–540.

[**Mand**] Mandelbaum, R., *Four-dimensional topology: an introduction*, Bull. A.M.S. **2** (1980), 1–159.

[**Mazur**] Mazur, B., *On embeddings of spheres*, Bull. A.M.S. **65** (1959), 59–65.

[**Melvin**] Melvin, P., *Four-dimensional oriented bordism*, in "4-Manifolds", Contemporary Math. **35** (1984), 399–406.

[**Milnor1**] Milnor, J., *On simply connected 4-manifolds*, Proc. Int. Symp. Algebraic Topology, Mexico (1958), 122–128.

[**Milnor2**] ________, *Spin structures on manifolds*, L'Ensignement Math. **8** (1962), 198–203.

[**Milnor3**] ________, *On manifolds homeomorphic to the 7-sphere*, Ann. Math. **64** (1956), 399–405.

[**Milnor4**] ________, *On the 3-dimensional Brieskorn manifolds* $M(p,q,r)$, in "Knots, Groups and 3-Manifolds", ed. L. Neuwirth, Ann. Math. St. **84** (Princeton University Press, 1975).

[**M-H**] Milnor, J. and Husemoller, D., "Symmetric Bilinear Forms," Springer-Verlag, Berlin and New York, 1973.

[**M-S**] Milnor, J. and Stasheff, J., "Characteristic Classes," Ann. Math. Studies **76** (Princeton University Press, 1974).

[**N-R**] Neumann, W. and Raymond, F., *Seifert manifolds, plumbing, μ-invariant, and orientation reversing maps*, Springer Lecture Notes in Math. **664** (1977), 163–196.

[**O-V**] Okonek, C. and Van de Ven, A., *Stable bundles and differentiable structure on certain elliptic surfaces*, Inven. Math. **86** (1986), 357–370.

[**Quinn1**] Quinn, F., *Ends of maps, III: dimensions 4 and 5*, J. Diff. Geom. **17** (1982), 502–521.

[**Quinn2**] ________, *Handlebodies and 2-complexes*, Springer Lecture Notes in Math. **1167** (1985), 245–259.

[**Rober**] Robertello, R. A., *An invariant of knot cobordism*, Comm. Pure and App. Math., XVIII (1965), 543–555.

[**Rohlin**] Rohlin, V., *New results in the theory of 4-dimensional manifolds*, Dokl. Akad. Nauk. S.S.S.R. **84** (1952), 221–224 (See also a French translation in [**G-M**].).

[**Rolfsen1**] Rolfsen, D., "Knots and Links," Publish or Perish, Boston, 1976.

[Rolfsen2] ________, *Rational surgery calculus: extension of Kirby's theorem*, Pacific Jour. Math. **110** (1984), 377–386.

[Rourke] Rourke, C. P., *A new proof that Ω_3 is zero*, J. London Math. Soc. (2) **31** (1985), 373–376.

[R-S] Rourke, C. P. and Sullivan, D. P., *On the Kervaire obstruction*, Ann. Math. **94** (1971), 397–413.

[Sch] Scharlemann, M., *The four-dimensional Schoenflies conjecture is true for genus two imbeddings*, Topology **23** (1984), 211–217.

[Serre] Serre, J. P., "Cours d'Arithmétique," P.U.F., Paris, 1970.

[Sieb1] Siebenmann, L. C., *Amorces de la chirurgie en dimension 4, un $S^3 \times R$ exotique (d'apres A. Casson et M. H. Freedman)*, Sem. Bourbaki **536** (1978–79).

[Sieb2] ________, *La conjecture de Poincaré topologique en dimension 4 (d'apres M. H. Freedman)*, Sem. Bourbaki **588** (1981–82).

[Stall] Stallings, J., *The piecewise linear structure of Euclidean space*, Math. Proc. Camb. Phil. Soc. **58** (1962), 481–488.

[Steenrod] Steenrod, N., "The Topology of Fiber Bundles," Prince University Press, 1951.

[Taubes] Taubes, C. H., *Gauge theory on asymptotically periodic 4-manifolds*, J. Diff. Geom. **25** (1987), 363–430.

[Trace1] Trace, B., *A class of 4-manifolds which have 2-spines*, Proc. A.M.S. **79** (1980), 155–156.

[Trace2] ________, *On attaching 3-handles to a 1-connected 4-manifold*, Pacific Jour. Math. **99** (1982), 175–181.

[Wall1] Wall, C. T. C., *Diffeomorphisms of 4-manifolds*, J. London Math. Soc. **39** (1964), 131–140.

[Wall2] ________, *On simply connected 4-manifolds*, J. London Math. Soc. **39** (1964), 141–149.

[Wall3] ________, *On the orthogonal groups of unimodular quadratic forms*, Math. Ann. **147** (1962), 328–338.

[Wall4] ________, *On the orthogonal groups of unimodular quadratic forms, II*, Crelle **213** (1963), 122–136.

[Wall5] ________, *Non-additivity of the signature*, Inven. Math. **7** (1969), 269–274.

[Wh1] Whitehead, J. H. C., *A certain open manifold whose group is unity*, Quart. J. Math. Oxford, ser. **6** (1935), 268–279.

[Wh2] ________, *On simply connected 4-dimensional polyhedra*, Comm. Math. Helv. **22** (1949), 48–92.

[Whitney] Whitney, H., *The self intersections of a smooth n-manifold in 2n-space*, Ann. Math. **45** (1944), 220–246.

INDEX

Lecture Notes in Mathematics

For information about Vols. 1–1296
please contact your bookseller or Springer-Verlag

Vol. 1297: Zhu Y.-l., Guo B.-y. (Eds.), Numerical Methods for Partial Differential Equations. Proceedings. XI, 244 pages. 1987.

Vol. 1298: J. Aguadé, R. Kane (Eds.), Algebraic Topology, Barcelona 1986. Proceedings. X, 255 pages. 1987.

Vol. 1299: S. Watanabe, Yu.V. Prokhorov (Eds.), Probability Theory and Mathematical Statistics. Proceedings, 1986. VIII, 589 pages. 1988.

Vol. 1300: G.B. Seligman, Constructions of Lie Algebras and their Modules. VI, 190 pages. 1988.

Vol. 1301: N. Schappacher, Periods of Hecke Characters. XV, 160 pages. 1988.

Vol. 1302: M. Cwikel, J. Peetre, Y. Sagher, H. Wallin (Eds.), Function Spaces and Applications. Proceedings, 1986. VI, 445 pages. 1988.

Vol. 1303: L. Accardi, W. von Waldenfels (Eds.), Quantum Probability and Applications III. Proceedings, 1987. VI, 373 pages. 1988.

Vol. 1304: F.Q. Gouvêa, Arithmetic of p-adic Modular Forms. VIII, 121 pages. 1988.

Vol. 1305: D.S. Lubinsky, E.B. Saff, Strong Asymptotics for Extremal Polynomials Associated with Weights on R. VII, 153 pages. 1988.

Vol. 1306: S.S. Chern (Ed.), Partial Differential Equations. Proceedings, 1986. VI, 294 pages. 1988.

Vol. 1307: T. Murai, A Real Variable Method for the Cauchy Transform, and Analytic Capacity. VIII, 133 pages. 1988.

Vol. 1308: P. Imkeller, Two-Parameter Martingales and Their Quadratic Variation. IV, 177 pages. 1988.

Vol. 1309: B. Fiedler, Global Bifurcation of Periodic Solutions with Symmetry. VIII, 144 pages. 1988.

Vol. 1310: O.A. Laudal, G. Pfister: Local Moduli and Singularities. V, 117 pages. 1988.

Vol. 1311: A. Holme, R. Speiser (Eds.), Algebraic Geometry, Sundance 1986. Proceedings. VI, 320 pages. 1988.

Vol. 1312: N.A. Shirokov, Analytic Functions Smooth up to the Boundary. III, 213 pages. 1988.

Vol. 1313: F. Colonius. Optimal Periodic Control. VI, 177 pages. 1988.

Vol. 1314: A. Futaki. Kähler-Einstein Metrics and Integral Invariants. IV, 140 pages. 1988.

Vol. 1315: R.A. McCoy, I. Ntantu, Topological Properties of Spaces of Continuous, Functions. IV, 124 pages. 1988.

Vol. 1316: H. Korezlioglu, A.S. Ustunel (Eds.), Stochastic Analysis and Related Topics. Proceedings, 1986. V, 371 pages. 1988.

Vol. 1317: J. Lindenstrauss, V.D. Milman (Eds.), Geometric Aspects of Functional Analysis. Seminar, 1986–87. VII, 289 pages. 1988.

Vol. 1318: Y. Felix (Ed.), Algebraic Topology – Rational Homotopy. Proceedings, 1986. VIII, 245 pages. 1988.

Vol. 1319: M. Vuorinen, Conformal Geometry and Quasiregular Mappings. XIX, 209 pages. 1988.

Vol. 1320: H. Jürgensen, G. Lallement, H.J. Weinert (Eds.), Semigroups, Theory and Applications. Proceedings, 1986. X, 416 pages. 1988.

Vol. 1321: J. Azéma, P.A. Meyer, M. Yor (Eds.), Séminaire de Probabilités XXII. Proceedings. IV, 600 pages. 1988.

Vol. 1322: M. Métivier, S. Watanabe (Eds.), Stochastic Analysis. Proceedings, 1987. VII, 197 pages. 1988.

Vol. 1323: D.R. Anderson, H.J. Munkholm, Boundedly Controlled Topology. XII, 309 pages. 1988.

Vol. 1324: F. Cardoso, D.G. de Figueiredo, R. Iório, O. Lopes (Eds.), Partial Differential Equations. Proceedings, 1986. VIII, 433 pages. 1988.

Vol. 1325: A. Truman, I.M. Davies (Eds.), Stochastic Mechanics and Stochastic Processes. Proceedings, 1986. V, 220 pages. 1988.

Vol. 1326: P.S. Landweber (Ed.), Elliptic Curves and Modular Forms in Algebraic Topology. Proceedings, 1986. V, 224 pages. 1988.

Vol. 1327: W. Bruns, U. Vetter, Determinantal Rings. VII, 236 pages. 1988.

Vol. 1328: J.L. Bueso, P. Jara, B. Torrecillas (Eds.), Ring Theory. Proceedings, 1986. IX, 331 pages. 1988.

Vol. 1329: M. Alfaro, J.S. Dehesa, F.J. Marcellan, J.L. Rubio de Francia, J. Vinuesa (Eds.): Orthogonal Polynomials and their Applications. Proceedings, 1986. XV, 334 pages. 1988.

Vol. 1330: A. Ambrosetti, F. Gori, R. Lucchetti (Eds.), Mathematical Economics. Montecatini Terme 1986. Seminar. VII, 137 pages. 1988.

Vol. 1331: R. Bamón, R. Labarca, J. Palis Jr. (Eds.), Dynamical Systems, Valparaiso 1986. Proceedings. VI, 250 pages. 1988.

Vol. 1332: E. Odell, H. Rosenthal (Eds.), Functional Analysis. Proceedings. 1986–87. V, 202 pages. 1988.

Vol. 1333: A.S. Kechris, D.A. Martin, J.R. Steel (Eds.), Cabal Seminar 81–85. Proceedings, 1981–85. V, 224 pages. 1988.

Vol. 1334: Yu.G. Borisovich, Yu.E. Gliklikh (Eds.), Global Analysis – Studies and Applications III. V, 331 pages. 1988.

Vol. 1335: F. Guillén, V. Navarro Aznar, P. Pascual-Gainza, F. Puerta, Hyperrésolutions cubiques et descente cohomologique. XII, 192 pages. 1988.

Vol. 1336: B. Helffer, Semi-Classical Analysis for the Schrödinger Operator and Applications. V, 107 pages. 1988.

Vol. 1337: E. Sernesi (Ed.), Theory of Moduli. Seminar, 1985. VIII, 232 pages. 1988.

Vol. 1338: A.B. Mingarelli, S.G. Halvorsen. Non-Oscillation Domains of Differential Equations with Two Parameters. XI, 109 pages. 1988.

Vol. 1339: T. Sunada (Ed.), Geometry and Analysis of Manifolds. Proceedings, 1987. IX, 277 pages. 1988.

Vol. 1340: S. Hildebrandt, D.S. Kinderlehrer, M. Miranda (Eds.), Calculus of Variations and Partial Differential Equations. Proceedings, 1986. IX, 301 pages. 1988.

Vol. 1341: M. Dauge, Elliptic Boundary Value Problems on Corner Domains. VIII. 259 pages. 1988.

Vol. 1342: J.C. Alexander (Ed.), Dynamical Systems. Proceedings, 1986–87. VIII, 726 pages. 1988.

Vol. 1343: H. Ulrich, Fixed Point Theory of Parametrized Equivariant Maps. VII, 147 pages. 1988.

Vol. 1344: J. Král, J. Lukes, J. Netuka, J. Vesely´ (Eds.), Potential Theory – Surveys and Problems. Proceedings, 1987. VIII, 271 pages. 1988.

Vol. 1345: X. Gomez-Mont, J. Seade, A. Verjovski (Eds.), Holomorphic Dynamics. Proceedings, 1986. VII. 321 pages. 1988.

Vol. 1346: O.Ya. Viro (Ed.), Topology and Geometry – Rohlin Seminar. XI, 581 pages. 1988.

Vol. 1347: C. Preston, Iterates of Piecewise Monotone Mappings on an Interval. V, 166 pages. 1988.

Vol. 1348: F. Borceux (Ed.), Categorical Algebra and its Applications. Proceedings, 1987. VIII, 375 pages. 1988.

Vol. 1349: E. Novak, Deterministic and Stochastic Error Bounds in Numerical Analysis. V, 113 pages. 1988.

Vol. 1350: U. Koschorke (Ed.), Differential Topology Proceedings, 1987, VI, 269 pages. 1988.

Vol. 1351: I. Laine, S. Rickman, T. Sorvali (Eds.), Complex Analysis, Joensuu 1987. Proceedings. XV, 378 pages. 1988.

Vol. 1352: L.L. Avramov, K.B. Tchakerian (Eds.), Algebra – Some Current Trends. Proceedings. 1986. IX, 240 Seiten. 1988.

Vol. 1353: R.S. Palais, Ch.-l. Teng, Critical Point Theory and Submanifold Geometry. X, 272 pages. 1988.

Vol. 1354: A. Gómez, F. Guerra, M.A. Jiménez, G. López (Eds.), Approximation and Optimization. Proceedings, 1987. VI, 280 pages. 1988.

Vol. 1355: J. Bokowski, B. Sturmfels, Computational Synthetic Geometry. V, 168 pages. 1989.

Vol. 1356: H. Volkmer, Multiparameter Eigenvalue Problems and Expansion Theorems. VI, 157 pages. 1988.

Vol. 1357: S. Hildebrandt, R. Leis (Eds.), Partial Differential Equations and Calculus of Variations. VI, 423 pages. 1988.

Vol. 1358: D. Mumford, The Red Book of Varieties and Schemes. V, 309 pages. 1988.

Vol. 1359: P. Eymard, J.-P. Pier (Eds.) Harmonic Analysis. Proceedings, 1987. VIII, 287 pages. 1988.

Vol. 1360: G. Anderson, C. Greengard (Eds.), Vortex Methods. Proceedings, 1987. V, 141 pages. 1988.

Vol. 1361: T. tom Dieck (Ed.), Algebraic Topology and Transformation Groups. Proceedings. 1987. VI, 298 pages. 1988.

Vol. 1362: P. Diaconis, D. Elworthy, H. Föllmer, E. Nelson, G.C. Papanicolaou, S.R.S. Varadhan. École d´ Été de Probabilités de Saint-Flour XV–XVII. 1985–87 Editor: P.L. Hennequin. V, 459 pages. 1988.

Vol. 1363: P.G. Casazza, T.J. Shura, Tsirelson´s Space. VIII, 204 pages. 1988.

Vol. 1364: R.R. Phelps, Convex Functions, Monotone Operators and Differentiability. IX, 115 pages. 1989.

Vol. 1365: M. Giaquinta (Ed.), Topics in Calculus of Variations. Seminar, 1987. X, 196 pages. 1989.

Vol. 1366: N. Levitt, Grassmannians and Gauss Maps in PL-Topology. V, 203 pages. 1989.

Vol. 1367: M. Knebusch, Weakly Semialgebraic Spaces. XX, 376 pages. 1989.

Vol. 1368: R. Hübl, Traces of Differential Forms and Hochschild Homology. III, 111 pages. 1989.

Vol. 1369: B. Jiang, Ch.-K. Peng, Z. Hou (Eds.), Differential Geometry and Topology. Proceedings, 1986–87. VI, 366 pages. 1989.

Vol. 1370: G. Carlsson, R.L. Cohen, H.R. Miller, D.C. Ravenel (Eds.), Algebraic Topology. Proceedings, 1986. IX, 456 pages. 1989.

Vol. 1371: S. Glaz, Commutative Coherent Rings. XI, 347 pages. 1989.

Vol. 1372: J. Azéma, P.A. Meyer, M. Yor (Eds.), Séminaire de Probabilités XXIII. Proceedings. IV, 583 pages. 1989.

Vol. 1373: G. Benkart, J.M. Osborn (Eds.), Lie Algebras. Madison 1987. Proceedings. V, 145 pages. 1989.

Vol. 1374: R.C. Kirby, The Topology of 4-Manifolds. VI, 108 pages. 1989.